Environment, Development, and Sustainability

Environment, Development, and Sustainability

Perspectives and cases from around the world

Edited by

Gordon Wilson

Pamela Furniss

Richard Kimbowa

Published by Oxford University Press, Great Clarendon Street, Oxford OX2 6DP
in association with The Open University, Walton Hill, Milton Keynes MK7 6AA.

 OXFORD
UNIVERSITY PRESS

Oxford University Press is a department of the University of Oxford.
It furthers the University's objective of excellence in research, scholarship,
and education by publishing worldwide in

Oxford New York

Auckland Cape Town Dar es Salaam Hong Kong Karachi
Kuala Lumpur Madrid Melbourne Mexico City Nairobi
New Delhi Shanghai Taipei Toronto

With offices in

Argentina Austria Brazil Chile Czech Republic France Greece
Guatemala Hungary Italy Japan Poland Portugal Singapore
South Korea Switzerland Thailand Turkey Ukraine Vietnam

Oxford is a registered trade mark of Oxford University Press
in the UK and in certain other countries

Published in the United States
by Oxford University Press Inc., New York

First published 2010

British Library Cataloguing in Publication Data
Data available

Library of Congress Cataloging in Publication Data

Environment, development, and sustainability : perspectives and cases from around the world / edited by
Gordon Wilson, Pam Furniss, Richard Kimbowa.
    p.   cm.
Includes bibliographical references.
ISBN: 978–0–19–956064–6
1. Sustainable development – – Case studies.   2. Environmentalism – – Case studies.
I. Wilson, Gordon, Dr.   II. Furniss, Pamela.   III. Kimbowa, Richard.
HC79.E5E57254 2010
333. 7 – – dc22

2009022552
Typeset by Macmillan Publishing Solutions
Printed in Italy
on acid-free paper by
L.E.G.O. S.p.A.

This book is a learning resource for Open University courses. Details of Open University courses
can be obtained from the Student Registration and Enquiry Service, The Open University, PO Box 197,
Milton Keynes MK7 6BJ, United Kingdom: tel. +44 (0)845 300 60 90, email general-enquiries@open.ac.uk
http://www.open.ac.uk

ISBN: 978–0–19–956064–6

1 3 5 7 9 10 8 6 4 2

In memory of Salome Kamuri Okayi from Uganda who epitomized the international partnership that has created this book, but who is unable to witness the fruit of her labours.

■ PREFACE

This is an ambitious book about a topic with huge implications for the whole of humanity. Just a scan through the titles of the 27 chapters illustrates something of its scale. Nevertheless, one must also note from the outset that a book such as this can never be exhaustive in its coverage.

The concept of the book was first considered during discussions within the Sustainable Development Network, a coalition of UK Open University academics and African practitioners that was formed in 2004. Our starting point was the assumption that, at whatever scale we operate, issues of environment, development, and sustainability are locked together and one cannot think of one element without thinking of the other two.

The concept took further shape at an international workshop in Kampala, Uganda, in 2006, organized by the network but attended by other interested parties. Like the Sustainable Development Network itself, it was agreed that the book would seek unity in diversity with respect to how people in different parts of the world and people of different professional mindsets relate to issues of environment, development, and sustainability. Through its authorship, it would bring together north, south, east and west; academics and practitioners; scientists and social scientists.

These discussions pointed towards a multi-authored 'Reader' which deliberately exposes the reader to a multiplicity of issues and perspectives, while suggesting the possibility of synthesis. While it was agreed, therefore, that chapters would be stand-alone so that it is possible to dip in and out of the book, it was also considered imperative that we try to make sense of the array of messages emanating from each chapter, and to bring them together within a uniting narrative. Thus, in addition to editorial-style chapters at the start and end of the book, we decided that there would be reviews of each section.

These basic ideas have remained. The book is a Reader framed by a narrative which uses the concepts of difference and interdependence to provide a measure of coherence and an exploration of the synthesis possibilities. These concepts cut across those of environment, development, and sustainability in the book title. Difference celebrates the range of perspectives and cases as a potential source of learning and new knowledge, while interdependence is a realization that we as human beings are mutually dependent on each other and with the biophysical world in which we live.

Thus, although *Environment, Development, and Sustainability* is written as an introductory text book for undergraduate students, and the chapters have been written in an accessible manner, at the level of the whole book the narrative is quite challenging by the end. As already noted, it is possible to read individual chapters on a 'need to' basis, but we hope that many of you will be interested in engaging with the book as a whole and its more challenging aspects.

The differences between the chapters do not only relate to the topics they cover. We have deliberately chosen a range of authors who, whatever the topic they are addressing, approach it with different perspectives. The following broad, yet overlapping, distinctions between author perspectives are apparent. Between:

- Authors from different parts of the world, for example between authors from Africa, Australia, Europe, and the United States. All of the countries featured in the book are highlighted on the world map which follows this Preface.
- Academic and practitioner authors.
- Scientists, social scientists, technologists, and others.

It is not the aim to make judgements on any one of these perspectives in favour of the others, but, in the overall spirit of the book, to treat each with respect and as having validity, while, again, suggesting the importance of synthesis between them.

This idea of both respecting difference of perspectives, while looking for a more holistic approach, distinguishes this book from the great number of published books that already exist on sustainable development. These other titles tend either to focus on particular topics (such as energy, water, biodiversity) or on a specific region, or come at the subject from separate disciplinary perspectives. In contrast, *Environment, Development, and Sustainability* represents an examination of these key words of the title as three integrated issues that form the main ongoing challenge of this century, respecting all, but privileging no particular disciplinary approach.

Also, unlike many other titles on sustainable development, this Reader does not focus per se on developing countries. Although the limiting developing/developed country dichotomy is in such common use that one can't seem to get away from it, this book sets out to reflect the fact that development is not a finished state. Yes, it does contain chapters which focus on what we generally consider to be the 'developing' world. This group includes China, Cuba, Ethiopia, Kenya, India, Niger, South Africa, Uganda, and Zimbabwe (although some of these countries are developing in an economic sense relatively rapidly—the 2009 world recession as the book goes to press not withstanding). The book also contains chapters, however, which focus on Australia, Eastern and Western Europe, and the United States. These are richer regions and countries which nevertheless continue to develop (and conversely have pockets of awful deprivation), and where environment, development, and sustainability are as much intertwined as elsewhere in the world.

A final note which relates to the above paragraph concerns terminology when referring to (and classifying) different countries in relation to their level of 'development'. The term 'Third World' was originally coined about half a century ago to describe countries newly independent from colonial rule who would form a collective political entity distinct from the 'First World' (the capitalist countries of Western Europe, North America, Australasia, and Japan), and the 'Second World' (the former Soviet Union and its satellites). These terms have largely been dropped today for a variety of reasons. The terms 'developing' and 'developed' countries are still in use, however, to denote poorer and richer countries respectively, but this distinction is criticized on the grounds outlined above that no country has stopped developing. In more common use today are the terms the 'global south' and 'global north' which make a similar distinction based on wealth, although they are not geographically accurate (e.g. some rich countries are in the southern hemisphere!). In this book, however, we have decided to ignore these debates about terminology and allow chapter authors to use which terms and distinctions they wish.

The structure of the book

The ordering of chapters has undergone many iterations. This is understandable for a book that has deliberately sought difference and contrasts. We could have ordered the chapters into sections based on major topics, such as climate change, biodiversity, waste management and environmental health, participatory practices and so on. This might have been neat and logical, although inevitably some chapters would not have lent themselves to such classification. Such a structure, however, would have tended to compartmentalize the chapters somewhat within the sections when we wanted a more dynamic interplay between them.

Instead, therefore, we have a structure which, within each section and between sections, gives the sense of a wide range of locally experienced issues, themes and possible actions, and where there is a sense of flow (albeit an imperfect flow) from one section to the next. This structure divides the book into five Parts.

Part A, Introduction, comprises only the editorial Chapter 1. It explores several of the issues hinted at in this Preface and starts the book narrative.

Part B, Learning from the local, comprises Chapters 2 to 9. It provides a taster of the range of issues and perspectives on environment, development, and sustainability across the world—China, Uganda, Niger, Ethiopia, Eastern Europe, Cuba, Australia, the United States, and India (Mumbai) all feature. In this Part we adopt a broad conception of 'local', ranging from whole countries to single cities to village communities.

Part C, Major themes in environment, development, and sustainability, comprises Chapters 10 to 16. Although places still form an intrinsic part of most of the chapters in this Part, the focus is on the big themes of climate change, biodiversity, conservation, water management, solid waste management, globalization, and ethics.

Part D, Action for environment, development, and sustainability, comprises Chapters 17 to 25. It covers the varied attempts by both public and private actors to manage power and conflict at country, city, and community scales, so that difference becomes a resource for collective knowledge, innovation, and policy.

As noted above, each of the substantive Parts B–D ends with an editorial 'review essay', thus providing a linking narrative for the whole book.

Part E is the book conclusion and comprises Chapters 26 and 27. Chapter 26 makes explicit the different, broad, discipline-based perspectives which are often implicit in previous chapters, and argues that each is necessary but none, by themselves, are sufficient to engage completely with the environment, development, and sustainability challenge. The final Chapter 27 attempts to integrate everything, by expanding the early ideas of linkage and difference that are espoused in Chapter 1 to the notion of interdependence.

One final point, before you start. You will find in many chapters key terms and concepts highlighted in bold type. These terms and concepts are further explained in a glossary at the end of the book.

■ FIGURE ACKNOWLEDGEMENTS

Thank you to all the contributors who have provided photos included in this book.

Figure 2.1 Central Intelligence Agency **Figure 2.2** © Jason Lee/REUTERS **Figure 3.1** National Environment Management Authority, Uganda **Figure 3.2** © Mulondo Ssenkaali **Figure 3.3** © Mulondo Ssenkaali **Figure 4.1** © Kerstin Danert **Figure 4.2** © Kerstin Danert **Figure 4.3** © Kerstin Danert **Figure 4.4** © Kerstin Danert **Figure 6.1** © A. J. Lloyd **Box 4.1 and 4.2 photographs** © Kerstin Danert. **Figure 6.3** © James Warren **Figure 7.1** © David Wicks/John Fairfax Publications **Figure 8.1** © USATourist.com **Figure 9.1** © Maps of India **Figure 9.2** © Nikhilesh Haval/World of Stock **Figure 9.3** Reproduced with kind permission from Jeff LaMarca **Figure 10.1** Figure 1.2 from http://www.hm-treasury.gov.uk/stern_review_report.htm. Reproduced under the terms of the Click-Use Licence **Figure 11.2a** © Chushkin/Fotolia.com **Figure 11.2b** © Warren Kovach/Alamy **Figure 14.1** © Stephen Burnley **Figure 15.1** © Koichi Kamoshida/Getty Images **Figure 16.1** International Rivers Network © Eureka Cartography, Berkeley, CA **Figure 16.2** © Narmada Bachao Andolan **Figure 18.1** © Greater London Authority **Figure 18.2** © Greater London Authority **Figure 19.1** Reproduced with kind permission from IITA **Figure 19.2** www.fromthefrontline.co.uk **Figure 20.1a** © Robin Ray **Figure 20.1b** © Richard Hearne **Figure 20.2** © Robin Ray **Figure 21.1** © A. J. Lloyd **Figure 21.2** © ARUP **Figure 21.3** © ARUP **Figure 22.1** © Environment Africa. **Figure 24.2** © Kevin Winter.

■ ACKNOWLEDGEMENTS

These acknowledgements are inadequate for two reasons. First, in a book of this breadth and complex logistics resulting from its international authorship, it is impossible to name everybody who has helped along the way. Secondly, words on a page can never express the gratitude that is truly felt.

That said, the editors must thank:

The Mathematics, Computing, and Technology Faculty of the Open University for finding a way of supporting the project;

The Co-publishing unit at the Open University, first Christianne Bailey and later Giles Clark and David Vince.

The Kulika Trust, an educational charity that sponsors Open University Development Management students in Uganda. The Kampala office of Kulika organized sessions, and the international workshop in the city where the book finally started to take shape.

Design consulting engineers, Arup, and the British Council in Uganda who supported financially the international workshop in Kampala.

The numerous people who contributed to the Sustainable Development Network, out of which the project began, and who attended the Kampala workshop, but who in the end were not able to author any of the chapters. Your contributions to shaping the book have nevertheless been highly valued.

Suzanne Brown who managed the Open University end of the project in its early years, and Roger Dobson who took over the reins later.

Finally, but by no means least, Marie Lacy who has been secretary to the project from the start, and who has truly been a tower of strength in seeing it through. In the final days of preparing for handover to the publisher she was ably assisted by Debbie Derbyshire and Sharon Lumbers. We could not have made it without you.

Gordon Wilson, Pamela Furniss, and Richard Kimbowa

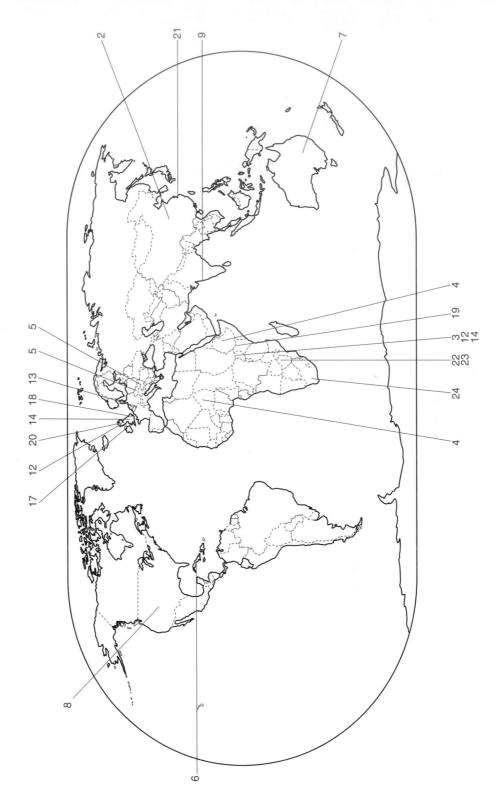

Where in the World? Map indicating locations featured in this book, with the relevant chapter number

■ CONTENTS

SECTION E CONCLUSION

■ AUTHOR BIOGRAPHIES

Dina Abbott is a University Reader in Development Geography at the University of Derby. She has researched extensively on urban poverty issues and slum livelihoods in Mumbai. Her recent research has focused around women's urban agricultural activities in west and east Africa.

Barbara Banda was born in Malawi and attended schools in Zimbabwe, Malawi, and the UK. Barbara has been a development practitioner for 25 years with experience in Southern Africa. Her passion is to see communities taking ownership of their own destinies. She has an M.Sc. in Development Management from the Open University.

Isaac Banadda Nswa is a chemist and waste management specialist working for Uganda Environmental Protection Forum (UEPF) since 1995. He is also a Consultant on environment and waste management (partnership building, development of local level policies and action plans, education and awareness material, and training needs assessment). His interest is in sustainable solid waste management approaches.

Chris Blackmore is a Senior Lecturer in Environment and Development at the Open University, involved in researching and teaching environmental decision making and systems since the early 1990s. She currently focuses on environmental and social learning systems and communities of practice. Chris has worked at national and international levels and written many book chapters, journal articles, and teaching texts.

Roger Blackmore is a Senior Lecturer and Staff Tutor at the Open University. His academic background is in Earth Sciences: Geography BA Oxford; Meteorology and Climatology, M.Sc. Birmingham; and scientific instrumentation, Photographic Technology B.Sc. Westminster. He has worked in education for 25 years teaching adults and school-leavers a variety of science, engineering and environment topics in further education colleges and universities. For the last ten years he has written open learning materials for a variety of Open University courses, including for the University's Environmental Studies degree. He currently chairs the production team for the University's first year environment course (U116).

Godfrey Boyle directs the Energy and Environment Research Unit at the Open University and is a visiting professor at the Energy and Resources Institute (TERI) University in Delhi. He has a 30-year track record of research and teaching in renewable and sustainable energy and has chaired a number of successful Open University courses in this field. He is co-author/editor of *Renewable Energy* (Oxford, 2004), *Energy Systems and Sustainability* (Oxford, 2003), and *Renewable Electricity and the Grid* (Earthscan, 2007). He is a Fellow of the Institution of Engineering and Technology, a Fellow of the Royal Society of Arts, and a Trustee of the National Energy Foundation.

Stephen Burnley is a Senior Lecturer in Environmental Engineering at the Open University. His interests cover teaching waste management at undergraduate and postgraduate level and researching the impact of solid waste management policy and legislation on practice in the community.

Joanna Chataway is Professor of Biotechnology and Development at the Open University. She is also the Open University Co-Director of the Economic Social Research Centre (ESRC) Innogen Centre. Innogen conducts interdisciplinary research into innovation in the life sciences and is shared with the University of Edinburgh.

David Cooke is a Senior Lecturer and Staff Tutor at the Open University. He is a Chartered Engineer, Fellow of the Energy Institute, and a Member of the Chartered Institution of Waste Management. He has experience of producing teaching materials, audio/visual and computer models in the environmental field at all undergraduate levels and of chairing both the production and presentation of distance teaching courses. His main field of interest is in computer modelling of waste management systems.

Ben Crow is an Associate Professor of Sociology at the University of California Santa Cruz. He previously taught at Stanford, Berkeley and the Open University. His current research is on how people in poor countries get access to safe and sufficient water. His books include *Markets, Class and Social Change, Sharing the Ganges* and the *Third World Atlas*.

Kerstin Danert is a water and sanitation specialist, with a focus on sub-Saharan Africa, working with governments, NGOs, and private enterprises. Her main interest is rural water supplies, particularly trying to reduce drilling costs, improve operation and maintenance of existing infrastructure, and introduce more affordable technologies.

Pamela Furniss is a Senior Lecturer in Environmental Systems at the Open University. She has many years experience of writing teaching materials for the university on a wide range of environmental subjects, principally in the areas of water resource management. She was Course Team Chair for the production of *Environmental decision making: a systems approach* (T863).

Kelly Gallagher Adjunct Lecturer in Public Policy, is Director of the Energy Technology Innovation Policy (ETIP) research group at the Harvard Kennedy School's Belfer Centre for Science and International Affairs. Her work is focused on studying, informing, and shaping US and Chinese energy and climate-change policy. She is the author of *China Shifts Gears: Automakers, Oil, Pollution, and Development* (The MIT Press, 2006).

Michael Gillman is an ecologist at the Open University with publications on a wide range of subjects, including conservation biology, plants, and butterflies of the Neotropics, evolution and mathematical models. He has a strong interest in interdisciplinary environmental teaching and research.

Charlene Hewat was born and brought up in Zimbabwe on a farm where her passion for the environment and community people was born. She cycled 22000 km from UK to Zimbabwe to raise funds and awareness for rhino conservation, after which she founded Environment Africa in 1990, an NGO which works in Southern Africa. Charlene has an M.Sc. in Development Management from the Open University.

David Humphreys is a Senior Lecturer in Environmental Policy at the Open University where he is interested in all areas of environmental governance. He has written two books on international forest politics, the second of which, *Logjam: Deforestation and the Crisis of Global Governance* (Earthscan, 2006), won the International Studies Association's 2008 Harold and Margaret Sprout Award.

Nicky Ison is an Australian youth climate activist. Nicky got her start as a climate activist at the age of 17 while completing the International Baccalaureate at Impington Village College just outside Cambridge, England. Since her return to Australia in 2004 she has been involved in the Australian Student Environment Network (ASEN). During her involvement in ASEN she has helped organize the annual Students of Sustainability Conference, was Environment Officer for the University of New South Wales Student Guild in 2006, ASEN National Convenor in 2007 and Climate Campaigner in 2008. Currently, she is finishing a double degree in Environmental Engineering and Arts at the University of New South Wales in Sydney, Australia. Nicky also works as a Research Consultant at the Institute of Sustainable Futures at the University of Technology, Sydney and hopes to get into the field of empowering communities through decentralized renewable energy projects.

Petr Jehlička is a Lecturer in Environmental Geography at the Open University. His work focuses on the development of environmental movements in central and eastern Europe and how domestic environmental traditions and discourses present under state-socialism combine with imported western know-how and agendas.

Raphael Kaplinsky is Professor of International Development at Development Policy and Practice at the Open University. In recent years he has played a leading role in global research and policy networks examining the operations of global value chains, the impact of China and India (the Asian Drivers) on other developing countries, and the impact of the commodities boom on development strategies.

Richard Kimbowa works for Uganda Coalition for Sustainable Development (UCSD), a local membership advocacy NGO. He is an alumnus of the OU's M.Sc. Development Management Programme (2003–06). He is professionally interested in civil society in terms of mechanisms and institutions to secure sustainable development.

Margaret Najjingo Mangheni (PhD) is a Senior Lecturer in the Department of Agricultural Extension/Education at Makerere University, Uganda. She has over 19 years of university teaching experience. She has also served as a consultant to international and national organizations in design, monitoring and evaluation of agricultural development programmes. She has written over 20 publications in refereed journals and edited books and proceedings of international professional meetings. Her current research interest is in the area of gender and other socio-economic crosscutting issues in agricultural development.

Adeline Muheebwa is a development practioner with over 13 years in active project-related activities in Uganda. She has an M.Sc. in Development Management from the Open University. She has previously worked with the UN and USAID-funded projects on increased agricultural production, environment protection, gender and energy conservation. She is currently working in a private company dealing in energy conservation cooking systems and environment protection strategies. Her passion is to promote wealth creation through partnerships while protecting the environment.

Fred Onyai worked in a decentralized local government system for 13 years implementing Government of Uganda policies, plans, and programmes/projects in environmental and natural resource management. Currently he is working with the National Environment Management Authority (NEMA) as the Internal Monitoring and Evaluation Specialist to ensure effectiveness and efficiency in environment management in Uganda. Besides, he lectures in Makerere University, Kampala in the Institute of Adult Education and Extramurals (Project Planning and Management). He has research interests in policy, programming, and process management with focus on environment management policies and projects, indigenous systems, and livelihoods approach to achieve sustainable development.

Martin Reynolds did his doctorate work at the Institute for Development Policy and Management (IDPM) at Manchester University based on the application of critical systems approaches to evaluating participatory approaches to natural resource management. A research post at the Centre for Systems Studies at the University of Hull generated a co-authored book *Environmental Management and Operational Research*. Since 2000, Martin has worked at the Open University producing distance learning course material for postgraduate courses on institutional development, environmental decision making, environmental responsibility, and systems approaches for managing change. He is one of the founding members of the Open Systems Research Group based at the OU, and provides workshop support for professional development in systems practice and critical systems thinking. He has published widely in the fields of systems studies, professional evaluation, international development, and environmental management (downloadable on Open Research Online http://oro.open.ac.uk).

Peter Robbins is a Senior Lecturer in Development Studies and Genomics at the Open University and Deputy Co-Director at the Open University of the ESRC Innogen Centre. He is author of *Greening the Corporation* (Earthscan, 2001), as well as articles on reflexivity in science and technology, including GM crops and food.

Robin Roy is Professor of Design and Environment at the Open University with a background in mechanical engineering, design, and planning. Since joining the Open University he has chaired and contributed to many distance teaching courses on design, technology, and the environment. He founded the Design Innovation Group to conduct research on product development, technological innovation, and sustainable design and is now co-chair of the Sustainable Technologies Group. He has published many books and papers on design, innovation, and environment.

Sandrine Simon. With a background in ecological economics, Dr Sandrine Simon initially focused on reforming economic policy tools to make them take better account of environmental considerations. She has also worked on participatory environmental decision making, the use of information and communication technologies in environmental education and policy, systems approaches to evaluation and environmental conflict resolution, sustainable water management and environmental responses from developing countries. She is a Lecturer in Environmental Systems at the Open University and contributes to courses in the Faculties of Mathematics, Computing and Technology, Social Sciences and Science.

James Smith is Co-Director of the Centre of African Studies in the Edinburgh International Development Centre, both at the University of Edinburgh. He is also responsible for coordinating developing country research at the Economic and Social Research Council Innovation and Genomics (Innogen) Research Centre.

Mulondo Ssenkaali is a development and humanitarian worker of 18 years with experience in management of development in Uganda, Sierra Leone, and Darfur in North Sudan. He is a graduate of the Pontifical Urbaniana University Rome (Philosophy), Development Studies Centre, Dublin, Ireland (Development), the Open University M.Sc. in Development Management and is currently completing the MA in Humanitarian Assistance at Tufts University, Boston, Massachusetts.

Alan Thomas is Visiting Professor at both the Centre for Development Studies, Swansea and the Open University, where he was instrumental in introducing undergraduate international development and the Global Masters in Development Management. He takes an action-oriented, case study approach to research, with interests in learning and knowledge sharing, development policy and management, and sustainable development.

James Warren is a Lecturer and Staff Tutor at the Open University, working on transport systems and modelling transport emissions and transport energy linked to the Design Group in the Faculty of Mathematics, Computing and Technology.

Roger Wheater has contributed to many aspects of conservation as a former Director of Uganda National Parks, Deputy Chairman of Scottish Natural Heritage and Chairman of the National Trust for Scotland. As Director of the Royal Zoological Society he was responsible for both national and international activities promoting species conservation, education, and research.

Gordon Wilson is a Senior Lecturer in Technology and Development at the Open University, where he is Director of the inter-faculty Environment, Development, and International Studies Programme. His research interests are in collective knowledge construction and the role of scientists, technologists, and other professional agents in development practice. He has undertaken research in Zimbabwe, Uganda, and Algeria. He is a Fellow of the Higher Education Academy.

Kevin Winter is a Lecturer in the Environmental and Geographical Science Department at the University of Cape Town. He is a lead researcher in the Urban Water Management Research Group based at the university. Research projects include studies in participatory action by the community in managing greywater and water demand management projects in the city of Cape Town.

SECTION A
Introduction

Making the connections between environment, development, and sustainability

Gordon Wilson

Introduction

The richest country in the world, the United States, also consumes the most energy per head of population. Although there were signs of change from a new President in 2009, hitherto the United States government had persistently refused to sign any international deals to limit its carbon emissions, on the grounds that to do so might jeopardize the country's economy and lifestyle.

Uganda in East Africa is one of the poorest countries in the world and consumes about ten times less energy per head of population than the United States. People who protest on environmental grounds against hydroelectric schemes on the River Nile and destruction of rainforest are invariably accused by the Government of being 'anti-development'.

These examples illustrate how *development*, which embraces the economy and society, and human-induced *environmental change* are intimately connected. The maxim is equally true of rich (more developed) and poor (less developed) countries. Moreover, human-induced environmental change arising from development has been mostly negative. It involves two processes—*depletion*, such as of oil resources, and *degradation* of air, land or water through pollution. The two are often connected: for example, pollution of a river by industry might poison fish, leading to their depletion.

Worldwide, there is no more pressing issue than this linkage. It was explicitly recognized in 1983, when the United Nations convened a World Commission on Environment and Development (WCED). The WCED itself was the culmination of well-publicized concern, which had originated in the 1960s over pollution, especially from pesticides (Carson 1962), and depletion of resources such as oil and the consequential limits on economic growth (e.g. Meadows *et al.* 1972).

Two world 'summits' on the same theme have followed the WCED—1992 in Rio de Janeiro and 2002 in Johannesburg, as well as several mini-summits on particular aspects. Not only is the linkage a major contemporary issue, therefore, it is also an enduring one. The rapid development and its impact on the environment in the world's most populous country, China, is currently concentrating minds still further, as Chapter 2 explores.

This book is founded on four basic premises. We have just outlined the first, which is that much human-induced environmental change is done in the name of development, or to maintain a developed state, wherever it occurs in the world, even if not always stated explicitly. This link between environment and development is therefore a major issue, possibly the major issue of our times.

A rough and ready view of *environment* associates it with 'nature'. This doesn't pass close examination, however, because so much that we consider as nature has been shaped to a significant extent by humans. Sometimes the shaping is deliberate, as in farming landscapes, sometimes it is unintentional through, for example, pollution of the air causing acid rain.

It is more instructive to think of our environment as the biophysical context in which we live and the source of our livelihoods. We should especially think of it as constantly changing under both human and non-human influences. Thus, Chapter 10 explains how our climate has always changed and continues to change.

The timescale of environmental change can be very long. It is, however, uneven and there have been periods of comparatively rapid change. Oil deposits take millions of years to form, yet human activity can use them up within a short time. With climate change, there are tipping points, such as periods when the earth has been plunged into an ice age. The current concern, as Chapter 10 elaborates, is that human activity is pushing towards another tipping point.

Development is explicitly about human change. One view of development is that, like environmental change, it is a long, uneven process, although now the timescale is definitely human and confined to the last few hundred years. A manifestation of development's unevenness is that the process does not produce benefits equally for the population. It tends to produce losers as well as winners. The present era of **globalization**—meaning the interconnectedness of the world, especially in terms of the economy—that is discussed in Chapter 15, can be seen as the latest phase in the historical process of development.

A second view of development, however, sees it simply as a vision of a good society towards which we strive, while a third sees it as deliberate interventions aimed at improvements in material and social conditions of our lives—such as interventions to provide clean water in rural areas of Africa that are considered in Chapter 4. See Thomas (2000) for elaboration on these views of development.

The three views are themselves interconnected. Deliberate intervention is designed usually to ameliorate the 'disordered faults' of the historical process which creates winners and losers (Cowen and Shenton 1996). It is also often designed with some kind of vision of a just society in mind. Sometimes it is designed with the aim of 'catching up' by a country that considers itself to be 'less developed' (e.g. many African countries) with those that are 'more developed' (e.g. the United States and other wealthy countries). Here the vision is 'modernization', as described in relation to agriculture in Uganda in Chapter 3.

The views of development also give rise to debate regarding what should be done about environmental change. Thus Chapter 8 notes that in the United States, one side elaborates the need for deliberate intervention to mitigate it, while the other claims that environmental challenges will be met within the tried and trusted process of capitalist development that the US epitomizes.

Environment, development, and sustainability

Sustainability summarizes attempts to meet the overall challenge presented by environment and development. Originally, in the 1950s, sustainability described the continued thriving of a biological species—such as a fish, bird or tree—which might become depleted or even extinct through human activity, over-fishing being a good example. Later it was extended to describe a whole ecosystem—for example, a forest with many interdependent plant and animal species—in similar terms. Conservation (Chapter 12) and biodiversity (Chapter 11) are directly linked to this notion of sustainability.

In the early use, one referred always to the sustainability *of* something, be it a particular fish or a tropical rainforest. A later use simply extended this to a broader set of 'sustainabilities': the sustainability of people's livelihoods, of a society's way of life (or culture), of a government's economic policy, of an international system for managing world trade, and so on. In this book, several chapters explore sustainability in relation to both narrower and broader uses of the term.

Three important features follow:

1. Sustainability is a *normative* concept, meaning that it expresses a desirable state—it is desirable for a species, and a human way of life to thrive; so too a country's economy, and its education and health systems.

2. Sustainability refers to the ability to continue over time, which requires robustness against shocks. But our use of the word 'thrive' above implies more. A group of peasants might be able to engage in a farming practice over time, but only in a degraded form as other pressures to gain livelihoods, such as taking temporary waged jobs for somebody else, means they don't sow at the right time, weed, or water their crops adequately. Sustainability therefore has two broad dimensions—robustness and effectiveness.

3. Different sustainabilities are interconnected. For example, the Stern Review (HM Treasury 2006) into climate change pointed out its economic costs and suggested that climate change is produced by an economic model of growth that will ultimately become unsustainable. Also, an economic model that means the rich become richer while the poor become poorer is likely to be socially unsustainable. These kinds of issues have led academics and practitioners to group sustainability issues under three broad headings or 'pillars of sustainability': environment, economy, and society.

We can summarize the above features in the book's second premise: sustainability is a desirable state that refers to the robustness of something and its continuing ability to do whatever it does effectively. There can be many 'sustainabilities', but it is usual to group them into three interconnected areas of concern: environment, society, and economy.

Because it relates to so much of the book, we end this subsection with a brief discussion of *sustainable development*. If we think of the three pillars of sustainability, development encompasses economy and society—the economy for our material well-being and society for our social well-being (and related elements such as well-being in terms of culture, health, education, and citizen rights). The concept of *sustainable development* connects economy and society to our environment.

There are well over 100 definitions of sustainable development in existence! The one most quoted, however, stems from the World Commission on Environment and Development (WCED) to which we referred above. The definition is often named after the Norwegian Prime Minister, Gro Harlem Brundtland, who chaired the WCED. It appeared in the WCED book, *Our Common Future*, as: 'development that meets the needs of the present without compromising the ability of future generations to meet their own needs' (Brundtland 1987: 43).

Since 1987, the definition has received rigorous scrutiny and refinement, hence the large number of alternatives which abound today. Most are based on the observation that the original definition concentrates on conserving the present for future generations (known as showing a concern for *intergenerational equity*) but ignores existing divisions in the world between, for example, rich and poor countries (a lack of concern for *intragenerational equity*). This led some writers to accuse the WCED of ultimately siding with nature and not justice (e.g. Sachs 1997: 294). The accusation has been a major source of debate between richer and poorer countries at the aforementioned Earth Summits, with poorer countries essentially demanding that:

- They too have a 'right to development'. The rich countries, therefore, should make available on concessionary terms new technologies that address environmental concerns while simultaneously enabling economic development.

- The rich countries should bear the cost of cleaning up the environmental mess because they created it, and continue to contribute to it in large share, through their past development.

Nevertheless, the Brundtland definition is the starting point for most of the other definitions, even if they end up looking rather different from it. It is also our starting point. You will see, in Chapter 17, just how difficult it is to nail down an agreed meaning of sustainable development, even in a very small country.

Creating sustainability

Almost everybody can agree upon the notion of sustainability as a desirable state. There are bound to be arguments, however, about what represents such a state, and how we might achieve it—the many definitions of sustainable development just noted being a symptom of these arguments. Nor is it necessarily a stable state. Here, it's important to take people, organizations, and countries as we are, in continual tension between how we would like the world to be and our more immediate interests. In European countries, for example, there is evidence that concern for the environment moves higher up the agenda when the economy is doing well, but when it is doing less well our concerns turn to the value of our houses and job security. In poor countries, clashes between what are perceived to be environmental needs and development needs are often very stark. Chapter 9 focuses on how, within a single city (Mumbai, India), rich and poor people view the environment in very different ways. This leads to the book's third premise: there are many different views on what constitutes sustainability in relation to environment and development and on how to achieve it.

Earlier I isolated 'robustness' and 'effectiveness' as two broad dimensions of sustainability. Robustness relates to the ability to continue over time, but it can also be taken to mean that sustainability is not necessarily a steady state. Robustness equates with the ability to adapt and evolve in line with changing contexts, and even shocks, over time. Sustainability is not, therefore, an end state, but is continually being reinvented.

The capacity among human beings to adapt and reinvent stems from our ability to *innovate*. **Innovation** is a term used in economics to describe the creation of new products that are sold. Here we articulate a more general definition: our ability to do new things in any sphere of action. Crucially, our ability to innovate depends on what we know—our knowledge—and a useful general definition of innovation is 'knowledge put to productive use' (Chataway 2005: 597). The centrality of innovation for sustainability is discussed in Chapters 19, 20, and 21.

We get to know through learning, but how do we learn? One powerful idea is that we learn both formally and informally from comparing what we don't know with what we already know. In short, we learn from difference. In this sense, it should be a cause for celebration that there are so many different views about environment, development, and sustainability, as they are resources for our learning and hence our knowledge and ability to innovate.

As with everything else, however, it's not so simple. Difference in the world is not often perceived positively as a rich source of learning. More likely it is associated with conflict between groups, defined ethnically, religiously, in terms of respective material wealth, by gender and so on. All too often such conflict is in violent form, which gives difference a negative connotation. More generally, such conflict is usually based on inequality between the groups and this is what difference really means. More generally still, *inequality* is itself a symptom of power relations between people and groups, and conflict is basically a struggle to readjust power relations. Knowledge itself is related to this power relation, with some people's knowledge being valued more than that of others (see for example the discussion of traditional/improved/modern water supply and irrigation in Niger and Ethiopia in Chapter 4). To give a crude comparison, the knowledge of how things are done in the United States is very powerful and serves as a model for many other countries. The knowledge of how things are done among a community in a tropical rainforest or a village in sub-Saharan Africa will be very different, however, and not valued as highly by outsiders. The situation is further complicated because the tropical rainforest community and the African villagers might themselves have internalized the idea that their knowledge is worth less than that which emanates from a United States citizen.

Where, then, does this leave us, if on one hand difference between people is a rich source of learning, while on the other it represents a power relation, inequality, and conflict? That is the ultimate challenge for this book, which has deliberately set out to provide you with different perspectives on environment, development, and sustainability in its chapters. These perspectives derive from the different knowledges of the authors. Some are based on their origins; others on their educational background (whether they are primarily natural scientists or technologists or social scientists—see Chapter 26); and still others on what they do, as academics, practitioners, and activists.

One practical challenge for the book is for us as editors to afford equal validity to each chapter. Another is to make the connections between them. This means going beyond celebrating the difference between our authors to generating new knowledge out of that difference, to make the whole greater than the sum of the individual parts. Although each chapter is self-contained and you may read only those chapters that are of direct interest to you, there is much to gain, therefore, from treating the book as an integrated whole.

A final challenge is to enable us (meaning 'you' the reader alongside 'us' the editors) to act in the real world, in whatever realm we find ourselves—as citizens, members of communities, professionals and lay people—with a sense of two things:

1. The power differences that circumscribe our relations to one another, whether as individuals, groups or whole countries.

2. The interconnectedness across scales, by which we mean the web of multi-directional influences between the global, national, and local scales.

Such action will vary from context to context, and emphasize different aspects. It might appear to be rather circumscribed and limited. But at least it stands a chance of being realistic, even if the demand feels like the impossible!

Much of the book is about action. Chapter 7 discusses environmental activism in Australia, while Chapter 16 examines how ethics can guide appropriate action, and Chapter 17 explores how it is negotiated politically in a small country (Wales). Chapters 19, 22, 23, and 24 explore 'participation' in a variety of settings—agricultural biotechnology in Kenya, industrial clusters and community development in Zimbabwe, and conservation in South Africa—as a practice for accommodating many voices in order to make decisions on what to do. Chapter 25 continues the theme by examining tools that might facilitate collective decision making, such as environmental impact assessment.

We now arrive at the book's fourth and final premise: the many perspectives on environment, development, and sustainability are a resource for us to learn from, gain knowledge and thereby, act appropriately. A major challenge, however, is to work within, while simultaneously challenging, the potentially negative dimensions of difference in terms of inequality and power relations.

SUMMARY

- This chapter has explored the key terms in the title: environment, development, and sustainability, and their interconnections through the concept of sustainable development.

- Out of this exploration, the chapter has developed four premises on which the book is based. Keep them in mind as you read the chapters which follow.

■ REFERENCES

Brundtland, G. (1987) *Our Common Future, Report of the World Commission on Environment and Development*. Oxford: Oxford University Press.

Carson, R. (1962) *Silent Spring*. Boston, MA: Houghton Mifflin.

Chataway, J. (2005) Introduction: is it possible to create pro-poor agriculture-related biotechnology? *Journal of International Development* 17(5): 597–610.

Cowen, M. and Shenton, R. (1996) *Doctrines of Development*. London: Routledge.

HM Treasury (2006) *Stern Review: The Economics of Climate Change, Executive Summary*. London: HM Treasury. Accessed online at http://www.hm-treasury.gov.uk/sternreview_index.htm

Meadows, D. H., Meadows, D. L., Randers, R. and Behrens, C. (1972) *The Limits to Growth*. New York: Universe Books.

Sachs, W. (1997) The need for the home perspective. In M. Rahnema and V. Rowntree (eds) *The Post-development Reader*, pp. 290–301. London: Zed Books.

Thomas, A. (2000) Meanings and views of development. In T. Allen and A. Thomas (eds) *Poverty and development into the 21st century*, pp. 23–50. Oxford: Oxford University Press.

SECTION B
Learning from the local

2

The challenge for environment, development, and sustainability in China

Kelly Gallagher

Introduction

China provides a unique laboratory to examine the intersection of environment, development, and sustainability. China is large in almost every dimension (Figure 2.1)—it has the world's biggest population, the second-largest economy by one measure, a big land area, and it is the world's largest consumer of coal. This chapter will show how energy is at the heart of the 'environment' problem, and environment is at the heart of

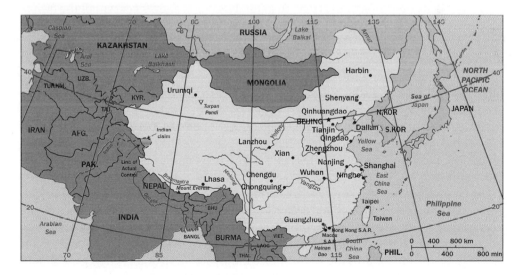

Figure 2.1 Map of the People's Republic of China.

the 'energy' problem in China. The chapter reviews the development and environment challenges facing contemporary China, and then examines the particular challenge of China's heavy reliance on coal for its energy needs. The chapter concludes with a discussion of what a more sustainable development path might look like in the Chinese context.

Development challenges in China

China's economic development during the past 30 years has been remarkable by nearly all metrics. During the 1970s, the Chinese economy was stagnant, and hundreds of millions of people languished in **poverty**. Since then, China has consistently been the most rapidly growing economy on earth, with average annual economic growth rates of 10 per cent since 1978 (Naughton 2007). Adjusted for differences in cost of living in different countries, known as **purchasing power parity (PPP)**, China is now the second-largest economy in the world (UNDP 2007/2008) as measured by **gross domestic product (GDP)**. Unadjusted per capita GDP (i.e. GDP per person) has grown from 1,595 yuan (US$230) to 17,100 yuan (US$2,500) during the past 30 years (Pan and Gallagher 2009), while on a PPP basis China's per capita income in 2005 was US$6,750 (HDI 2008). As a result of this steady and rapid economic growth, an estimated 400 million people have been pulled out of absolute poverty since 1979 according to the World Bank (2003).

Despite these achievements, the Chinese government continues to face difficult development challenges. First, inequality is widening dramatically, which could cause instability and political unrest. Inequality is most stark between urban and rural areas, and especially between western and eastern coastal provinces. As of 2002, the incomes of urban residents were three times larger than rural incomes, and 99 per cent of the people living in absolute poverty were living in rural areas, especially in the western provinces. Infant mortality was three times higher in rural areas than in urban ones (World Bank 2003). The 'Gini coefficient' measure of inequality (where perfect equality = 0 and total inequality = 1) for China increased from 0.20 in the early 1980s to 0.469 in 2004, the latest year data are available. China now ranks 81st in the world in terms of economic inequality, below Korea, most of the Baltic states, Russia, and Brazil (UNDP 2007/2008). This inequality makes the Chinese government nervous about social stability, because as Susan Shirk (2007: 54) notes, 'The Communist Party considers rapid economic growth a political imperative because it is the only way to prevent massive unemployment and labor unrest.'

The centre-piece of the government's current five-year plan is to promote the service industries—the so-called 'tertiary' sector—because of their higher value added to the economy, and the energy and environmental benefits associated from a weaker reliance on heavy manufacturing. The current goal is to move the economy away from heavy industry and towards the service-based industries. In so doing, energy use should decline, and environmental quality should improve.

Environmental challenges in China

China's population is the largest in the world, at 1.3 billion in 2008. One in every five people on earth lives in China. Although the vast majority of these still live in rural areas, it is estimated that about 18 million Chinese migrate to cities each year (UNFPA 2007).

A major environmental problem in China is the lack of availability of fresh water. Two-thirds of China's cities lack sufficient water resources, and 110 cities suffer severe water shortages. An estimated 600 million people drink contaminated water daily (Economy 2004). The inefficient agricultural sector is mainly to blame for over-use, though industries are not blameless either. The Chinese government estimates 75 per cent of the river water that flows through urban areas is unfit for human consumption (Economy 2007). One-third of China's cities do not have waste water treatment plants (CCICED 2007) and, according to a recent survey by the State Environmental Protection Administration, half of the waste water treatment plants do not work properly (China Economic Review 2008).

China houses some of the largest hydropower installations in the world, most infamously the Three Gorges Dam. These hydropower stations have more localized but no less severe environmental impacts, forcing human relocation, inundation of agricultural lands, and impacts on biodiversity. On the other hand, hydropower projects can help with flood control, irrigation, and in the provision of low carbon-emission electric power. High carbon-emission coal-fired power plants are also the fourth-largest source of waste water in the country (Sinton *et al.* 2000).

In China's biggest cities most of the urban air pollution comes from motor vehicles. The car population in China has grown dramatically, from fewer than 100,000 in 1990 to approximately 25 million in 2007. While the growth in new cars has been astounding, the total number is still small compared with the situation in the United States, which has a private vehicle population of 230 million. With 20 per cent of the world's population, the Chinese only own 1.5 per cent of the cars in the world (Davis and Diegel 2007).

Chemical wastes in China are not well controlled, and numerous instances of accidental or deliberate releases of toxic wastes have been documented. Notably, the Songhua river in the north-east has suffered repeatedly from xylidine and nitrobenzene releases from factories, both of which are highly dangerous to human health. As with other environmental problems in China, lack of enforcement of existing environmental laws related to managing chemical and other toxic wastes (including electronic equipment wastes) is a major challenge to sustainability.

Energy use and air pollution

China's energy-related challenges are many, including the need for energy to sustain economic growth, its increasing foreign dependency for oil and gas, the need to provide modern forms of energy to China's poor, its increasingly severe urban air pollution, its already massive acid deposition (commonly known as acid rain), the growing concerns domestically and internationally about global climate change, and access to affordable, advanced energy technologies to address all of the above challenges. Simply, energy is at the heart of the environmental problem, and environment is at the heart of the energy problem (Holdren 2008).

Total energy consumption in China increased 70 per cent between 2000 and 2005, and total coal consumption increased by 75 per cent during the same time period (World Bank and SEPA 2007). This astonishing rate of growth indicates that China's entire energy system is doubling in size every five years so far this century. Because of China's large population, even if everyone consumed a very small amount of energy, China's total energy consumption would still be large.

Economically, China's growing energy consumption presents both challenge and opportunity. Energy sustains economic growth by providing fuel for factory boilers, electricity for lighting and machinery, and for transportation services. Of course, energy provides heating and cooling services as well. During periods of energy shortages in China, factories are shut down entirely, or moved into a rotation where each factory must cease production periodically—once a week, or every other night. During the summer of 2004, for example, factories on the outskirts of Shanghai were forced to close two days a week due to the high demand for electricity to power air conditioners and insufficient supply (Fallon 2004). More recently, heavy snowstorms in central China in January 2008 contributed to severe electricity shortages in 31 provinces, caused primarily by an acute shortage of coal. This shortage was exacerbated by several factors, namely strong and persistent demand for coal, the closing down of small, unsafe coal mines, rising coal prices, and price controls on electricity. Because the central government sets electricity prices, as coal prices rise but electricity prices stay the same, profit margins shrink until power producers are forced to shut down. As a result of the coal shortages, the Chinese government suspended coal exports for two months, causing coal prices to shoot up to all-time highs in the United States, Europe, and Asia. China's largest copper producer, Jiangxi Copper Co., shut down some plants due to the high coal prices and lack of availability, as did some steel, zinc, and aluminum producers (Oster and Davis 2008).

The economic costs of China's air pollution are high. According to a recent report from the Chinese government and World Bank, conservative estimates of morbidity and premature mortality associated with ambient air pollution in China was equivalent to 3.8 per cent of GDP in 2003 if premature death was valued at 1 million yuan per person. Acid rain, caused mainly from sulfur dioxide (SO_2) emissions from coal combustion, is estimated to cost 30 billion yuan in crop damage (mostly for vegetables) and 7 billion yuan in material damage annually. This damage is equivalent to 1.8 per cent of the value of the crop output.

Providing better energy services to the poor in order to improve the quality of life for those still reliant on traditional forms of energy such as charcoal, crop wastes, and dung is a preoccupation of the Chinese government as part of its socio-economic development strategy. Nationally, 96 per cent of rural households have access to electricity, although in some provinces the figure is much lower, such as in Guizhou where 80 per cent of households lack access (LBNL 2001).

One rising concern is that as China imports greater amounts of energy, prices of these commodities will rise until supply catches up, and price spikes are especially likely during supply disruption events. After China became a net oil importer in 1994, its demand for global oil supplies grew rapidly as it became the second-largest consumer of oil in 2004, and it is now the third-largest oil importer in the world (BP 2008). China's escalating demand contributed strongly to the rise in world oil prices during the first part of this century as suppliers scrambled to catch up. It appears that China is beginning to affect world coal prices as well. In the first six months of 2007, China imported more coal than it exported for the first time in history. Overall, China's coal demand grew nearly 9 per cent in 2007, indicating that Chinese coal demand could double by 2015.

The other big challenge is to supply enough energy, especially electricity, to meet the very high demand created by Chinese industry. Industry consumes the majority of electricity supplied, accounting for 74 per cent of total demand (NBS 2006). On the opportunity side, the Chinese energy sector represents an exciting market opportunity for Chinese and foreign energy services companies alike. In 2006, China installed 101 gigawatts (GW) of new power plants, 90 GW of which was coal-fired power. In 2007, China installed an additional 91 GW for a total of 713 GW (China Electricity Council 2008). To put these numbers in perspective, Germany's entire electricity system in 2005 was 124 GW (Eurostat 2008).

Just in the last decade, China has emerged as a major consumer of oil, and there is strong potential for China to become a major natural gas consumer as well, especially if it tries to reduce its greenhouse gas emissions (natural gas is much less carbon-intensive than coal). About half of China's oil imports come from the Middle East, but Angola became the largest single supplier in 2006, and indeed, China has invested heavily in energy resources in Africa. Although there have been several new oil discoveries in China recently, Chinese reserves are on the decline. China has relatively few natural gas reserves domestically, and therefore uses virtually no natural gas in its power sector. To offset this, it is trying to increase production of coal-bed methane, a form of natural gas extracted from coal deposits. If China decides to increase its reliance on natural gas, it will have to import. If it begins to import large quantities from either the Persian Gulf or Russia, the geopolitical implications would be significant. China's long-term energy security is not only dependent on having sufficient supplies of energy to sustain its economic growth, but it will be equally dependent on its ability to manage the growth in energy demand without causing intolerable environmental damage.

The particular challenge of coal

China's main energy resource endowment is coal (Figure 2.2). Coal accounts for 93 per cent of China's remaining fossil fuel resources. In China, 74 per cent of the electricity is derived from coal, at 526 GW in 2007. Hydropower provides 20 per cent of electricity capacity, nuclear 1 per cent, and wind power half of one per cent. Although nuclear and wind power have been growing rapidly in recent years, coal is so dominant that it is unlikely that the current mix of electricity supply can be significantly altered any time soon. Natural gas is not commonly used for power generation due to the high price and lack of availability of the fuel (see above). China is aggressively pursuing renewable energy, and ranks number one in the world in some respects, such as in its installation of solar hot water and small hydropower. It ranks fifth in the world in terms of installed wind

Figure 2.2 China's coal-dependent economy.

capacity, and fourth in terms of ethanol production (Martinot 2007). Still, China's non-hydro renewable capacity is a tiny fraction of primary energy supply, which is completely dominated by coal.

Coal is at the heart of China's environmental woes, with major implications for human health. Particulate matter from coal is a major air pollutant. Average concentrations of PM10 (particles the size of 10 microns or less that are capable of penetrating deep into the lung) in China's cities are extremely high, ranging from (in parts per million) 255 in Panzhihua, 150 in Beijing, 140 in Chongqing and 100 in Shanghai. These numbers compare with 45 in Los Angeles and 25 in New York. PM10 can increase the number and severity of asthma attacks, cause or aggravate bronchitis and other lung diseases, and reduce the body's ability to fight infections. Certain types of people are especially vulnerable to PM10's adverse health effects, including children, the elderly, exercising adults, and those suffering from asthma or bronchitis (CARB 2003). In addition, each year, more than 4,000 miners die in China's coal mines, mostly due to accidents (Biallo 2008).

Sulfur dioxide emissions from coal combustion, a major source of acid deposition, rose 27 per cent between 2001 and 2005. Acid rain predominantly affects south-eastern China, and Hebei Province is most severely affected, with acid rain accounting for more than 20 per cent of crop losses (World Bank and SEPA 2007).

In terms of global climate change, coal emits the most greenhouse gas emissions of any fossil fuel and accounts for approximately 80 per cent of China's carbon dioxide (CO_2) emissions. The possible adverse impacts of climate change on China are likely to concern:

- Water supply and agriculture due to changing rainfall patterns and disappearing glaciers. For example, precipitation decreased by 50–120 mm/year along the northern Yellow River between 1956 and 2000, an already arid region. Conversely, precipitation increased by 60–130 mm/year along the southern Yangtze River from 1956–2000, an area that has been plagued by heavy flooding. The mountain glaciers on the Tibetan plateau are receding rapidly, with major implications for fresh water supply in already water-stressed northern China.

- Sea levels, where a rise of 30 cm would cause massive coastal inundation, which Chinese scientists estimate would cost 115–120 billion yuan in economic losses to the major delta areas of the Pearl, Yangtze, and Yellow Rivers, including the Gulf of Bohai.

Achieving sustainable development in China

It is challenging to imagine China (and most other countries) achieving true environmental sustainability, if one defines a sustainable process or condition as 'one that can be maintained indefinitely without progressive diminution of valued qualities inside or outside the system where the process operates or the condition prevails' (Holdren, Daily and Ehrlich 1995: 3). But it is not, in fact, difficult to imagine environmental conditions being vastly improved in China, nor is it hard to imagine China formulating a new mode of industrialization that is far cleaner and more efficient than the US model, for example. Indeed, one could even envision a future where the Chinese government decided to embark on a completely new growth strategy that championed sustainable development precisely because the environmental woes currently afflicting China are so severe and costly to its society.

To achieve dramatic environmental improvements in China, a comprehensive and far-reaching incentive system would have to be created. Experience to date suggests that even though much more energy-efficient and cleaner technologies have been developed, sometimes within China, they are not widely adopted in the country. The proposition that late-industrializing countries like China would 'leapfrog' to the most sustainable technologies available has not been borne out, and in fact, many limits to leapfrogging have been identified. Most importantly, it is clear that the processes of leapfrogging, technology transfer and cooperation, and accelerated deployment of environmental technologies are not automatic (Ohshita and Ortolano 2002; Gallagher 2006; Lewis and Wiser 2007).

Technology transfer partly concerns the transfer of products, machinery and processes from one (typically richer) country to another (typically poorer) country, sometimes through investment by a large foreign firm in the country and sometimes through development aid. Crucially, to be successful it also requires transfer of skills, and maintenance and management capabilities, so that the receiving country can develop its own technological capacity over time.

Technological leapfrogging refers to the possibility of poorer countries adopting cleaner, greener technologies without first having to go through the stage of employing dirty technologies as the now-rich countries did during their development.

Many barriers to technological leapfrogging exist in different contexts, including the higher costs of some cleaner technologies, lack of knowledge about or access to those technologies, and insufficient incentives to adopt them. Indeed, without clear and consistent incentives for firms to produce and consumers to purchase cleaner products and services, they often fail to do so. Especially because environmental quality is a public good, government has a special role to play to design and enforce environmental laws and regulations, which in turn create the appropriate incentives for producers and consumers alike.

The Chinese government recognizes and is tackling many of China's environmental problems. The government, for example, has issued strong targets for energy efficiency and renewable energy. The target for renewable energy is 15 per cent of primary energy by 2020 (Martinot 2007). In addition, it has issued six main environmental laws and nine natural resources laws, while The State Council has released 28 environmental administrative regulations, and the Ministry of Environmental Protection has published 27 environmental standards. Reportedly, more than 900 local environmental rules have been promulgated as well (Liu 2007). Many of these laws and regulations are somewhat weaker than their counterparts in the United States and Europe, but some are actually stricter or more far reaching. China's passenger car fuel-efficiency standards, for example, are considerably more stringent than those approved by the US Congress in 2007.

On the other hand, enforcement of China's environmental policies is highly uneven. Some cities like Beijing in its run-up to the 2008 Olympics went to tremendous lengths to clean up their local factories and reduce air and water pollution. Generally, local environmental enforcement is lax and undermines the relatively good policies that have been issued by the central government. For its part, the central government has thus far failed to provide adequate resources to strengthen the Ministry of Environmental Protection. There is no adequate system of environmental data collection, distribution, and analysis, which further complicates the enforcement effort because without irrefutable data about pollutant emissions and effluent releases, the government lacks a fundamental tool that would enable it judge and act upon non-compliance with the law. All of these deficiencies demonstrate the need for more effective government institutions to promulgate and enforce regulations.

In Europe and the United States, the environmental movement was and continues to be critically important to the passage and enforcement of landmark environmental laws. The formation of public interest groups or non-governmental organizations (NGOs) like the World Wildlife Fund, Greenpeace, Natural Resource Defense Council, and Sierra Club as well as 'green' parties created new and powerful political forces. These parties and NGOs advocated for environmental protection by educating citizens, pushing for new laws, monitoring enforcement, and suing firms or government agencies for non-compliance. In Japan, a culture of conservation sprang from the realization during the Second World War that it is a country with large resource needs but few endowments of natural resources.

In China, environmental groups are allowed to form, but usually only for the purposes of public education. A somewhat bizarre form of NGOs called 'government-owned non-governmental organizations' (GONGOs) initially emerged. More 'pure' environmental NGOs now exist, but they are still mostly confined to public education activities. The government has apparently given the media permission to report on environmental abuses, and it has established 'hotlines' for citizens to call to report environmental infractions. Still, it is clear that criticism of government policies, and especially the Communist Party itself, is not acceptable. Average citizens and NGOs are not potent political forces with respect to the formation of environmental policies today in China. There is, however, a growing reliance on academia to inform environmental policy making where university and research institute experts are encouraged to make suggestions, recommendations, and even relatively modest constructive criticisms to the government. It is difficult to imagine that China would be able to forge a sustainable path without the help of non-governmental organizations and institutions given the large size of its economy and population.

The final crucial component of a transition to an environmentally sustainable future in China is money. Large financial resources will be needed because even though China is already a successful industrializing country, there are competing demands for available financial resources. If the rest of the world wants China to move more quickly to reduce its environmental impact on the planet, then other countries will almost certainly need to provide greater financial resources to help. Money will also be needed for investment in technological innovation. Investments in the research, development, demonstration, and deployment of more environmentally sustainable technologies will reduce the costs of these technologies in the near term and provide a better menu of options for the future.

■ SUMMARY

- The challenge for environment, development, and sustainability in China is enormous because of China's large population, heavy reliance on coal, and China's energy intensive economy.

- Coal is at the heart of China's environmental woes, with major implications for human health.

- In terms of the causes of global climate change, coal emits the most greenhouse gas emissions of any fossil fuel. Coal accounts for approximately 80 per cent of China's carbon dioxide (CO_2) emissions.

- To achieve dramatic environmental improvements in China, a comprehensive and far-reaching incentive system would have to be created.

- Government has a special role to play to design and enforce environmental laws and regulations, which in turn create the appropriate incentives for producers and consumers alike.

- The Chinese government recognizes many of China's environmental problems.

- There is a need for need for more effective government institutions to promulgate and enforce regulations in China.

■ REFERENCES

Biallo, D. (2008) Can coal and clean air co-exist in China? *Scientific American*, August 4, http://www.sciam.com/article.cfm?Id=can-coal-and-clean-air-coexist-china.

BP (2008) *Statistical Review of World Energy*. BP, available from http://www.bp.com/statistical review, June.

CARB (2003) *Air Pollution – Particulate Matter Brochure*. Sacramento: California Air Resources Board, Sacramento.

CCICED (2007) *Global Experiences and China's Solution. Environmentally Sound Chemicals Management in China*. CCICED Issues Paper. Accessed 09 January 2009, http://www.vancouver.sfu.ca/dlam/.

China Economic Review (2008) *China's Environment 2008*. Hong Kong: China Economic Review Publishing.

China Electricity Council (2008) *Nation-Wide Power Sector Statistics Newsletter for 2007*. (In Chinese), http://www.cec.org.cn.

Davis, S. C. and Diegel, S. W. (2007) *Transportation Energy Data Book. Report ORNL-6978, Oak Ridge National Laboratory*. Oak Ridge, TN: US Department of Energy.

Economy, E. (2007) The great leap backwards? *Foreign Policy* 86(5): 38–59.

Eurostat (2008) *Infrastructure—electricity—annual data*. Eurostat. Downloaded from epp.eurostat. ec.europa.eu.

Fallon, B. (2004) Energy shortage hits Chinese firms. BBC News, http://news.bbc.co.uk/2/hi/business/3602678.stm, downloaded 8 February 2008.

Gallagher, K. S. (2006) Limits to leapfrogging? Evidence from China's automobile industry. *Energy Policy* (34): 383–394.

HDI (2008) *Human Development Index, Fact Sheet for China*. Available from http://hdrstats.undp.org/countries/country_fact_sheets/cty_fs_CHN.html.

Holdren, J. P., Daily, G. and Ehrlich, P. (1995) The meaning of sustainability: biogeophysical aspects. In M. Munasinghe and W. Shearer (eds) *Defining and Measuring Sustainability The Biogeophysical Foundations*, pp 3–17. Washington, DC: World Bank.

Holdren. J. P. (2008) Science and technology for sustainable well-being. *Science* 319: 424–434.

LBNL (2001) *China Energy Data Book*. Berkeley, CA: Lawrence Berkeley National Laboratory, Version 5.0.

Lewis, J. I. and Wiser, R. H. (2007) Fostering a renewable energy technology industry: an international comparison of wind industry policy support mechanisms. *Energy Policy* 35: 1844–1857.

Liu, X. (2007) Building an environmentally-friendly society through innovation: challenges and choices. Background Paper, China Council for International Cooperation on Environment and Development.

Martinot, E. (2007) *Renewables 2007 Global Status Report*, REN21. Paris: Renewable Energy Policy Network.

Naughton, B. (2007) *The Chinese Economy: Transitions and Growth*. Cambridge, MA: The MIT Press.

NBS (2006) *Statistical Yearbook of 2006*, Chapter 7. Washington, DC: China National Bureau of Statistics.

Ohshita, S. B. and Ortolano, L. (2002) The promise and pitfalls of Japanese cleaner coal technology transfer to China. *International Journal of Technology Transfer and Commercialisation* 1(1 and 2): 56–81.

Oster, S. and Davis, A. (2008) China spurs coal-price surge. *The Wall Street Journal*, 12 February, A1.

Pan, J. and Gallagher, K. S. (2000) Global warming: the road to restraint. In G. Allison, G. Guoliang, R. Rosecrance, C. H. Tung and J. Wang (eds), *Power and Restraint: A Shared Vision of the US China Relationship*, pp. 119–136. New York: Public Affairs.

Shirk, S. (2007) *China: Fragile Superpower*. Oxford University Press: New York.

Sinton, J. E., Fridley, D. G., Logan, J., Yuan, G., Bangcheng W., and Xu, Q. (2000) *Valuation of the Environmental Impacts of Energy Use in China*. Washington, DC: World Resources Institute.

UNDP (2007/2008) *Human Development Report: Fighting Climate Change: Human Solidarity in a Divided World*. New York: United Nations Development Programme.

UNFPA (2007) *State of the World's Population.* New York: UN Population Fund.

World Bank (2003) *China: Promoting Growth With Equity.* Country Economic Memorandum, Report No. 24169-CHA, October 15. Washington, DC: World Bank.

World Bank and SEPA (2007) *Cost of Pollution in China: Economic Estimates of Physical Damages.* Washington, DC: World Bank and China State Environmental Protection Administration.

▥ FURTHER READING

China's Policies and Actions for Addressing Climate Change. White paper issued by the Chinese Government, 29 October 2008. Available from http:/www.english.gov.cn/2008–10/29/content_1134544.htm.

World Bank and China State Environmental Protection Administration (2007) *Cost of Pollution in China: Economic Estimates of Physical Damages.* Washington, DC: World Bank and China State Environmental Protection Administration, 2007.

Elizabeth Economy (2007) The great leap backwards? *Foreign Affairs* 86(5): 38–59.

Kelly Sims Gallagher (2006) *China Shifts Gears: Automakers, Oil, Pollution, and Development.* Cambridge, MA: The MIT Press.

3

Rural development and environment in Uganda

Margaret Najjingo Mangheni, Mulondo Ssenkaali, and Fred Onyai

Introduction

Uganda is a low income country in sub-Saharan Africa whose economy is predominantly agrarian, the agricultural sector accounting for 36 per cent of **gross domestic product**, 81 per cent of the employed labour force, and 31 per cent of export earnings (Encyclopaedia of the Nations 2007). The sector, consisting of smallholder farmers cultivating an average of about 2.5 acres each, has been historically largely subsistence where farmers produce primarily for their own household consumption. In 2007, about 70 per cent of the area under cultivation was still used to produce locally consumed food crops.

However, Uganda is shifting towards commercial agricultural production and many previously subsistence crops are now increasingly grown for cash (Sorensen 1996). This expansion of commercial agriculture and other livelihood activities within a context of increasing population, poverty, poor governance, weak institutions, and socio-economic inequity has led to over-exploitation of natural resources (notably forest lands and fish stocks), environmental degradation, and social conflicts. The high population growth rate (2.5 per cent annually) alone is increasing demand for land, food, and energy (Uganda Government 2001).

Thus commercialization has created a tension between development and environmental sustainability. This chapter presents two case studies (see Figure 3.1) which depict this tension:

- Western Uganda where the social, livelihood, and environmental impacts of commercialization of cooking bananas have led to gender-based conflict and forest degradation with consequent loss of biodiversity;
- The Ssese Islands on Lake Victoria where economic development has led to natural resource degradation, especially of forests, fish stocks, and soils.

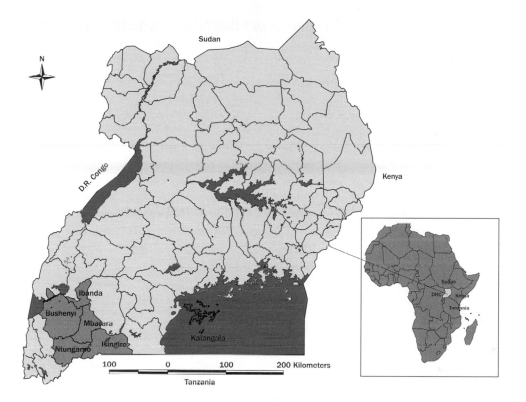

Figure 3.1 A map of Uganda and the two case study areas (shaded in blue)

In analysing these case studies, the chapter emphasizes the role of **governance, institutions** and organizations in explaining the environment and development challenges facing Uganda.

Governance is defined here as the conditions for ordered rule (Stoker 2004), involving a variety of stakeholders, in this case the government, the private sector, non-governmental organizations (NGOs) and local communities among others. Institutions are the humanly devised constraints that structure human interaction. They include formal constraints such as constitutions and laws and informal constraints such as norms, conventions and self-imposed codes of conduct. Organizations are groups of individuals bound together by some common purpose to achieve certain objectives. Generally, institutions define the rules of the game while organizations are the players (North 1990), although sometimes large influential organizations are also institutions—for example the government of Uganda.

As is often the case for a low-income country, Uganda has weak national and local organizations and institutions. Without governance mechanisms to negotiate the institutional rules of the game across the range of stakeholders, and organizations able to implement them, the desired balance between environment and development is hard to achieve. Despite the existence of various government policy instruments, engagement of certain key stakeholders in the process has been weak with negative consequences for the balance between environment and development.

Commercialization of banana production in south-western Uganda

Banana growing in Uganda dates back 2000 years (Karugaba and Kimaru 1999). Three kinds of bananas are grown: green cooking bananas, locally known as matooke, are cooked to use as a staple food. Beer bananas are a variety which is used to brew alcohol. Yellow sweet bananas are eaten uncooked when ripe. In its recent history, between 1970 and 1990, green cooking banana production rapidly declined in the traditional growing areas in the central region, while during the same period production quadrupled in the southwest (Gold *et al.* 1999) transforming the crop from subsistence to commercial (Steven and Breth 2000). Increased domestic demand resulting from a growing urban population in central and south-western Uganda, coupled with the fall of world market prices for Uganda's traditional export crops, especially coffee, is believed to explain this transformation (Steven and Breth 2000).

Commercialization of cooking bananas is associated with:

- higher investment in agricultural chemical inputs,
- use of hired labour,
- changes in production methods, including planting cooking bananas in pure stands as opposed to the traditional intercropping with other food crops and trees,
- practising soil and water conservation, and
- use of special tools to perform specific tasks such as desuckering, weeding, and detrashing.

In combination, these steps have improved productivity (Tumusiime forthcoming). Meanwhile, urban demand has driven crop expansion, usually at the expense of forests and wood lots (localized groups of trees often used for firewood). It has also displaced other food and cash crops due to scarcity of agricultural labour and shortage of land. As a result, commercialization of bananas has had impacts on **biodiversity** (especially 'agro-biodiversity'), cropping patterns, household harmony and food security, with implications for environmental and livelihood sustainability.

Effect on agro-biodiversity

Agro-biodiversity (i.e. agricultural biodiversity) refers to the diversity of cultivated crops and crop varieties, and livestock species and breeds within species in a farming system (Nkwiine and Tumuhairwe 2004). Farmers in south-western Uganda used to grow a diversity of crops and crop varieties for subsistence and for sale. However, the shift from subsistence to commercial farming has led to specialization in order to exploit economies of scale and maximize market-driven agricultural production (ibid.). Consequently there has been a negative impact on agro-biodiversity. In Isingiro District of south-western Uganda, for instance, the range of crops grown has decreased significantly with some crops such as beer bananas, groundnuts, peas, millet, Irish potatoes, sweet potatoes, cassava, and sorghum being abandoned completely by many farmers in favour of production of commercial cooking bananas (Tumusiime forthcoming).

The land devoted to these bananas has also encroached on what used to be public wood lots and forests. On occasion, this has been compensated by increased tree-planting elsewhere, as Tumusiime (ibid.) has reported in Nyamuyanja parish in Isingiro District, where each household should have its own source of fire wood in the form of wood lots. However, this has only been possible among the few richer households with enough land.

Effect on cropping patterns

The modern practices have changed the traditional cropping patterns in the area. Previously, for example, the system involved inter-cropping with bananas, coffee, and several other crops planted together to improve fertility and food availability, and also trees which provide fuel wood, windbreaks, and soil stabilization (Zake *et al.* 1998). With modern banana plantation practices, there has been a widespread removal of all inter-planted trees and other crops, removing these complementary benefits.

Effects on household harmony and food security

Traditionally, ensuring household food availability in south-western Uganda was the responsibility of women while men were engaged in cash crop enterprises and off-farm income-generating activities. Thus, women historically controlled production of cooking bananas since it was a food crop. However, commercialization of the crop has brought it under the control of men, resulting in gender conflict. For example, in Isingiro District, in several cases where the banana plantations had previously produced mainly at subsistence level, women were responsible for maintenance. However, once a woman had improved the husbandry practices and production had been raised to levels that generated substantial amounts of cash income, the husband would take over both the managerial and the marketing responsibility of the banana enterprise (Tumusiime forthcoming). The proceeds would then often be invested by the husband in small business or other ventures. Money earned from the sales was often not used to buy food and men tended to sanction only family consumption of poor-quality rejects of the commercial crop. In addition, growing of other food crops by women tended to be pushed to marginal lands, preference being given to cooking bananas for the fertile land.

All of this has impacted negatively on household food availability and increased domestic violence against women due to quarrels over money from the sale of bananas. In some instances the violence and quarrels have led to divorce. It is also ironic that south-western Uganda, which is referred to as the major 'food basket' for the country and a leading producer of cooking bananas (PMA 2001), still experiences household food insecurity and high levels of malnutrition, with the highest levels in the country of growth retardation among children below five years of age (USAID 2004).

Balancing development and environment in the Ssese islands

The 84 Ssese islands are located in the north-western corner of Lake Victoria in southern Uganda, with a total land area of 432 km² (Figure 3.2). Lake Victoria is the world's second-largest freshwater body, with an area of 67,850 km². Forty-three islands are

Figure 3.2 Tropical rainforest in the Ssese islands.

inhabited with a total population estimated at 50–60,000. Bugala is the largest island of the group and has the only sizeable settlement, Kalangala. Most of the islands have until recently been covered in tropical rainforest surrounded by unspoiled waters of the lake.

The local population gains its livelihoods mainly from fishing and subsistence agriculture. In recent years, however, lumbering has gradually expanded and some parts of the population have taken on charcoal burning. Overall, the population is poor in comparison with mainland Uganda where more economic opportunities are available. Literacy levels are low due to few schools and opportunities and social services such as health care and water provision are poor.

In addition to the ongoing poverty, there is now increasing concern about the potentially harmful changes that are taking place in fish stocks and forest-related resources, which we describe below.

Fish stocks

Lake Victoria was home to a large number of fish species, many of which are now thought to have been fished to extinction. The species still available in the lake in significant numbers are tilapia, Nile perch, silver fish and occasionally, mud and lung fish (Kalangala District 2005: 48–50). Catches of these fish are also dwindling rapidly, however, due to the increasing number of fishermen and the use of unsustainable means of fishing which sometimes includes poison. Due to increasing hardship on the mainland, many people are drawn to the islands by stories of big incomes to be made from fishing. The reality in the fishing sector is much less rosy. To increase catches, some fishermen use methods that catch fish in breeding areas, catch young fish, or simply harvest more fish than can be replaced through natural breeding. This is a sure way to fish to extinction and is threatening the livelihoods of the people of the islands, many of which can no longer be sustained through fishing.

Forest-related resources

The thick forest which until recently covered the islands is disappearing rapidly. There are three main reasons for this:

1. almost uncontrolled lumbering (by both the islanders and people from the mainland);
2. uncontrolled harvesting for charcoal burning and firewood (especially for fish smoking);
3. more recently, a large government-sponsored, vegetable oil development project.

Lumbering on the islands is driven by the near exhaustion of forests on mainland Uganda. Formerly, the choicest trees (mature, straight, and higher value trees such as nkoba or Uganda walnut) were cut but today all types and sizes are being felled. Boatloads are taken to the mainland and then to Uganda's capital, Kampala, every day. Also, in the face of loss of other forests on the mainland where it used to be produced, the Ssese islands are rapidly being turned into sources of charcoal. Much of the fish caught in the lake is not marketed fresh, due in part to transport problems, and is smoked before being transported to mainland markets. Whole islands, especially outlying islands, have lost their tree cover as a result of lumbering and charcoal burning.

The Vegetable Oil Development Project (Figure 3.3) is a partnership investment between the Ugandan government and a number of national and international vegetable oil producing and trading companies. An agreement worth US$120 million was signed in 2003 between the Uganda government and the Biwot Development Company (BIDCO) which is the leading implementer. As an international trading company with Ugandan and Kenyan subsidiaries, and linkages with companies in South Asia and elsewhere, BIDCO has the capacity to organize the palm oil production. The venture is funded by the United Nations International Fund for Agricultural Development (IFAD) and started in 2004. It involves planting 10,000 hectares of palm trees on Bugala island, covering about one-third of the island.

> The International Fund for Agricultural Development (IFAD) is a United Nations agency, established in 1977 to provide low-interest loans to combat rural hunger and poverty in developing countries (http://www.ifad.org).

Figure 3.3 The Vegetable Oil Development Project in Bugala Island: clearing land for the project.

Some of the densest forest on Bugala has been cleared to make way for the plantation. As a result, during the rainy seasons large amounts of exposed earth are carried away and deposited in the lake. There have been reportedly large numbers of dead fish floating in the lake in the affected areas. The clearance represents a huge impact on the forest resources of the country. With all of this forest destroyed, pressure will shift to the remaining forest on the other islands.

The Vegetable Oil Development Project does, however, arguably represent significant potential for improving the lives of the population in Ssese through employment and other livelihood opportunities arising from this new commercial crop grown at scale. It is claimed that a planned US$10 million processing plant to complement the plantation will not only provide local employment, but will also contribute to the overall economy of the islands and the country.

However, the project is problematic as it has not considered the environmental damage involved. The government, while creating a conducive economic environment for private sector investment, has allowed environmental destruction of a valuable physical resource. The presumption is that the environment is dispensable and its protection a luxury for a poor country like Uganda. Only when Uganda has attained a high level of economic development will the government start considering what to do about the environment (see also Chapter 12 for further discussion of these issues with respect to Mabira Forest on mainland Uganda).

However, simply believing and arguing that environmental conservation has to give way to a national economic development imperative is dangerous policy. In the long term abusing the environment has an economic cost too and will compromise whatever gains such an investment makes.

Balancing environment and development in Uganda

The two case studies concern substantially different contexts, although forest depletion is common to both. Throughout Uganda, however, the same overall choice is often presented, whatever the specific context: *development or the environment*. Is this really the case or is a third alternative possible?

The government of Uganda has an environmental policy framework which appears supportive of environmental protection, seeking to balance it with development needs. Environmental protection is enshrined in the 1995 constitution, and numerous auxiliary instruments and institutions are in place.

For forest protection, instruments include the National Environmental Statute (1995), the National Forestry Management Policy (2001) and the National Forestry and Tree Planting Act (2003). The government of Uganda has also established a programme of public sector reforms to improve on forestry management in the country. These reforms include decentralization of forestry management services with the establishment of District Forestry Services and lease of land in Forest Reserves to private tree farmers, and sector-wide approaches which involve integration of tree farming in other sectors notably agriculture. In Uganda, central government forest reserves are managed by the National Forestry Authority while the local government forest reserves are managed by District Forestry Services. Private forest reserves, however, which account for 70 per cent of the country's forest cover, are managed and regulated by the individual owners.

The government has also signed and ratified international conventions to protect forests and ensure sustainable use and management of forestry resources and products. The National Forestry Management Policy emphasizes an integrated forestry sector which seeks to achieve a sustainable increase in the economic, social, and environmental benefits from forests and trees for all Ugandans, especially the poor and vulnerable.

Attempts have also been made to control fish depletion. Policies have been formulated and legislation passed. Legislation is also sometimes implemented: for example, selling small young fish is banned, and patrols confiscate and burn nets that do not meet regulation sizes, while officials destroy unlicensed fishing canoes.

For agricultural commercialization, the subject of the first case study, the government has an overarching Plan for the Modernization of Agriculture (PMA) which aims to transform agriculture from subsistence to commercial production. It emphasizes the importance of product quality and quantity, and productivity gains—all of which are conventional economic concerns. It also emphasizes, however, sustainable use and management of the environment and natural resource base. The PMA therefore makes provisions for integrating environment considerations in its interventions.

All of this, however, has not succeeded in protecting the environment. The various laws, regulations, standards, and policy guidelines have been largely ineffective. Our two cases depict policy implementation failure and lack of adherence to the existing legal framework. A range of factors explain this situation, but key are weak institutions in the sense that they cannot ensure compliance to the rules of the game. Thus, formal rules cannot be enforced because of under-funding and informal rules regarding the environment/development balance are either not in place or are ignored by stakeholders as irrelevant to their activities. This situation is particularly challenging when the government does not even control most of the forest lands (see above).

Underlying weak institutions is the failure to include a range of stakeholders and facilitate an accommodation of their diverse interests. Such an accommodation would enable the stakeholders to negotiate the institutional rules of the game, both formal and informal, which they are prepared to accept as legitimate. The desired situation would be to arrive at a win–win position that all can live with. This is essentially a governance issue in the sense referred to in the introduction to this chapter.

For the Ssese islands, the most critical stakeholders include government and its departments; private interests engaged in charcoal burning, fishing, agriculture, and lumbering; local inhabitants; and NGOs who wish to conserve the environment. Here, however, no broader stakeholder engagement has occurred and in its absence the more powerful private interests have tended to dominate because they are able to persuade government to sometimes override its own instruments. Buchholz (1993) has observed that laws made without community and local popular participation are ineffective in sustainable management of environmental resources.

In the case of commercialization of cooking bananas in south-west Uganda, key stakeholders include government, local farmers (men and women), traders, and landless labourers on the commercial farms. In this case, weak enforcement of regulations is largely explained by poor institutional support to farmers to invest in more environmentally friendly practices. The poor farmers operating under circumstances of vulnerable livelihoods cannot be blamed for making the rational decision to respond to economic opportunities that meet their immediate needs as opposed to long-term environmental sustainability concerns. This case, however, additionally illustrates that the challenge is not simply to balance environmental concerns with those of economic development. Commercialization has also led to women losing access to land and ownership of production of an important crop, while men have gained. Inequitable social development is always a further complicating factor for economic development. Moreover, it is generally true that those who lose socially and economically through such development are also the ones who suffer most directly from harmful environmental impacts.

■ SUMMARY

- Commercialization of cooking bananas in south-western Uganda has had socially inequitable consequences alongside environmental degradation.

- In the Ssese Islands, fish stocks are declining while the rain forest is being depleted because of extensive lumbering, charcoal production, and a large government-backed vegetable oil plantation. Environmental degradation and adverse medium-term impacts on local livelihoods go hand in hand, although the vegetable oil project does potentially offer employment opportunities.

- It is argued that a balance has to be struck between environmental and economic development concerns. At present it is weighted too heavily towards economic development with not enough regard for the environment and its social impacts.
- This unhealthy balance is a function of:
  - (a) Poor governance, especially failure to bring on board and accommodate all the relevant stakeholders
  - (b) Weak institutions and organizations to negotiate the formal and informal rules of the game, and implement them.

▥ REFERENCES

Buchholz, A. R. (1993) *Principles of Environmental Management, The Greening Business*. Englewood Cliffs, NJ: Prentice Hall.

Encyclopedia of the Nations (2007) Available from http://www.nationsencyclopedia.com/africa/Uganda, accessed 28 March 2007.

Gold C. S., Karamura, E. B., Kiggundu, A., Bagamba, F. and Abera, A. M. K. (eds) (1999) Monograph on Geographic shifts in highland cooking banana (Musa, group AAA-EA) production in Uganda. *African Crop Science Journal* 7(3): 223–298.

Kalangala District (2005) *Kalangala District Development Plan 2005–6*. Uganda: Kalangala District.

Karugaba, A. and Kimaru, G. (1999) *Banana Production in Uganda: An Essential Food Crop and Cash Crop*. Nairobi, Kenya: Regional Land Management Unit (RELMA) and Swedish International Development Cooperation Agency (Sida). Available for download from http://www.worldagroforestry.org/sites/rela/Relmapublications/default.asp, accessed 30 March 2009.

Nkwiine C. and Tumuhairwe, J. K. (2004) Effect of market-oriented agriculture on selected agro biodiversity. Household income and food security components. *Uganda Journal of Agricultural Sciences* 9. 680–687.

North, D. C. (1990) *Institutions, Institutional Change, and Economic Performance*. New York: Cambridge University Press.

PMA (2001) *Eradication of Poverty in Uganda: Government Strategy and Operational Framework. Final Draft*. Entebbe: MAAIF.

Sorensen, P. (1996) Commercialization of food crops in Busoga, Uganda and the renegotiation of gender. *Gender and Society* 10(5): 608–628.

Steven, A. and Breth, S. A. (eds) (2000) *Food Security in Changing Africa*. Geneva: CASIN.

Stoker, G. (2004) *Transforming Local Governance*. Basingstoke: Palgrave Macmillan.

Tumusiime, B. (forthcoming) The effects of commercialization of cooking bananas (plantains) on household food availability in Isingiro district. MSc thesis, Makerere University, Department of Agricultural Extension/Education.

Uganda Government (2001). *The State of Environment Report for 2001/2*. Kampala: Ministry of Lands, Water and Environment.

USAID (2004) *Uganda Economic Development, Economic Growth, Agriculture and Trade*. Available from http://www.usaid.gov/policy/budget/cb2005/afr/pdf.

Zake J. Y. K, Nkwiine, C., Sessanga, J. K., Tumuhairwe, J. K., Bwamiki, D. Okwakol, M. Kasenge, V., Zobisch, M. A. and Sendiwanyo (1998) The effect of different land clearing and soil management practices on soil productivity in the Lake Victoria crescent zone of Uganda. Proceedings of the 16th Conference of the Soil Science Society of East Africa (SSSEA), Kampala, Uganda: Makerere University, Department of Soil Science pp. 293–299.

4

Traditional and modern (or improved) rural water supplies: stories from Ethiopia and Niger

Kerstin Danert

Introduction

The slogan 'water is life' is a truism which has become well known through the campaigns of various organizations. A person cannot survive for more than a few days if they do not drink. Crops will not root, grow, and flourish without water to nourish them. This chapter provides the reader with stories regarding the interaction between people, technology, and water in two of Africa's poorest nations and draws primarily on research and consultancy work by the author in these countries. These are stories of rural farmers, Non-Governmental Organizations (NGOs), and governments finding innovative ways of developing and sustaining rural water supplies and show that the distinction between 'traditional' and 'modern', or 'improved' and 'unimproved' is not always so clear.

Ethiopia and Niger—the former perched on the horn of Africa, the latter a landlocked expanse in the Sahel—are two countries commonly associated with poverty, hunger, and famine. If you watched the scenes broadcast on TV from Ethiopia in 1984, I doubt that you have forgotten them. Niger may have captured your attention too, with the recent drought in 2005.

Let me give you some of the statistics: Niger is currently ranked the sixth-poorest country in the world, with a human development index (HDI) of 0.370. Ethiopia, with its HDI of 0.389, holds the position of eleventh-poorest country (UNDP 2008).

> The HDI provides, on a scale of 0–1, a composite measure of three dimensions of human development: living a long and healthy life (measured by life expectancy), being educated (measured by adult literacy and enrolment at the primary, secondary and tertiary level) and having a decent material standard of living (measured by PPP [**purchasing power parity**] income).

HDIs for all countries of the World have been reported by the United Nations since 1990. Rich countries in North America, Western Europe, East Asia and Australasia all have an HDI greater than 0.9. Poor countries often have an HDI of less than 0.5. UNDP (2006)

Under-five mortality rates in Niger and Ethiopia for 2006 were respectively 253 and 123 children per 1000 live births (UNICEF 2008). Niger is home to an estimated population of 14 million with an annual growth rate of 3.3 per cent, while Ethiopia has a population of 77 million, growing at 2.4 per cent annually (UNFPA 2008). Both countries are highly dependent on subsistence agriculture (i.e. supporting oneself through direct consumption of farming produce rather than selling cash crops), with 83 and 84 per cent of the population living in rural areas in Niger and Ethiopia respectively. In Niger, an estimated 61 per cent of all people survive on less than the international absolute poverty line of US$ 1 per day; in Ethiopia it is 23 per cent (UNFPA 2008).

Inadequate access to water for domestic use and for watering crops and livestock is one of the major challenges facing the populations of these two countries. In 2004, only 36 per cent of rural dwellers in Niger were estimated to have access to 'improved' water supplies. The comparable figure for Ethiopia was 11 per cent (UNICEF/WHO 2008). The United Nations Food and Agriculture Organization (FAO) estimates that in Niger and Ethiopia respectively only 6 and 2 per cent of water from renewable resources was withdrawn in 2000, indicating a very low use of irrigation (FAO 2008).

Despite their abject poverty, both Ethiopia and Niger have some encouraging stories to tell with respect to innovative ways to improve domestic and agricultural water supplies.

'Traditional' and 'modern': never the two shall meet

Over the years, professionals around the world have polarized the concept of domestic water supply into *improved* and *unimproved* sources. In a similar vein, small-scale irrigation tends to be considered as *traditional* or *modern*. These distinctions are generally considered as black and white (Table 4.1). Although perhaps useful from a statistical point of view, as shown by the data presented above, they leave out a range of possibilities in between. Further, the black and white definitions tend to focus the attention of practitioners on specific water supply technologies rather than on the users, their struggles, aspirations, and ideas.

The improved or modern technologies tend to be favoured by professionals, practitioners, and political leaders. It is these technologies that enable the boxes to be ticked as we strive to achieve the United Nations **Millennium Development Goals** (MDGs), and specifically the MDG that aims to halve the proportion of people without access to safe water supplies between 1990 and 2015. Improved or modern technologies are relatively expensive and thus externally financed (by donors, NGOs, or government). Water users in sub-Saharan Africa are rarely encouraged to make their own incremental improvements to their domestic or agricultural water supply. As a result, they end up waiting for external assistance which may never be forthcoming. However, with some basic skills,

Table 4.1 Characteristics of improved and modern water supplies versus unimproved and traditional.

Improved domestic water supplies	Unimproved domestic water supplies
Household connection or public standpipe Borehole with pump Protected spring Rainwater collection	Unprotected spring Untreated water from a river; stream or lake Open well Puddle on road surface
Modern small-scale irrigation	Traditional small-scale irrigation
Gabion (wire boxes filled with rocks and stones) and concrete diversion structures Water applied by sprinkler or drip	River diversions of brushwood and stones Unlined canals

Sources: Carter and Danert (2006); Cotton and Bartram (2008: 19).

tools and a little capital many people could render their existing water supplies cleaner, more reliable and more efficient.

The other important point to note is that *drinking* water and water for *productive use* (e.g. watering animals or crops) tend to be considered by different government ministries, departments, NGOs, and professionals. In contrast, the rural dweller has a much more holistic way of conceptualizing his or her water.

Hand-drilled wells and water lifting in Niger

In much of Niger, as in many other parts of West Africa, hand-dug wells have been used as water sources for generations. Traditionally, these water sources were unlined, or supported with wood (Figure 4.1). Development projects have introduced modern wells, lined with bricks or mortar to improve safety and water quality. In Niger an open, cement-lined well is considered as modern, whereas in many other countries (such as Uganda), a well is only considered modern if it is covered and installed with a hand pump.

The majority of Niger's population depends on rain-fed agriculture, with small-scale irrigation undertaken on a limited scale, primarily in the south of the country. The market gardeners traditionally use calabashes (a water container made from a locally harvested large woody gourd) and leather bailers (in the form of a leather bag) to lift water from hand-dug wells and pour onto their crops. However, over the course of 30 years, improved water lifting devices have been introduced and taken up by farmers (Box 4.1), and small diameter hand-drilled wells have complemented the dug wells. Hand augering involves drilling a small-diameter borehole with a soil auger (Figure 4.2). This drilling method can penetrate certain sands and silts as well as some clay formations.

Richard Koegel of the Food and Agriculture Organization (FAO) introduced manual drilling techniques as early as the 1960s. Initiatives by the Peace Corps (an American volunteer agency) were followed by a Lutheran World Relief (LWR) initiative in the late 1980s. By 1996, it was estimated that the number of hand-drilled wells in southern Niger had grown to an estimated 3,500 (Naugle 1996) and now there are well over 18,000

Figure 4.1 A traditional hand dug well.

hand-drilled wells in use. Three major projects have substantially driven the adoption of hand-drilled well and treadle pump technology in Niger:

- The Tarka Valley Project in Madaoua region (1992 to 1997).
- The Pilot Private Irrigation Project (PPIP I) in the regions of Tilaberi, Dosso, Maradi, and Zinder (1997 to 2001).
- The Private Irrigation Project (PIP II) from 2002 to 2007.

The foundation for the adoption of these technologies was laid by earlier, independent projects that helped develop metalwork skills and established workshops in rural Niger. Commercialization of the technologies began in 2001, when the implementing organization of PPIP I (by the US NGO Enterprise Works/VITA) provided drillers with equipment on credit for constructing demonstration wells. Treadle pumps were manufactured and sold by local workshops. Building up a market in products that were at the time unknown to farmers was a challenge. Initially, the technique was to demonstrate the technology to users, ensure that they were able to use it properly, and then sell it to them. Links between drilling enterprises, pump manufacturers, and farmers were established immediately to avoid over-dependence on external project staff.

Hand-drilling of wells for agriculture has subsequently spilled over into the domestic market. Not only do farmers drink from their 'irrigation wells', but enterprises have also been

---

BOX 4.1 CALABASHES AND TREADLE PUMPS: CHANGES IN IRRIGATION PRACTICES

Sabiou Boubou (top right) uses his calabash to lift water. He raises the smooth round vessel from his hand-dug well, and in one swift movement places it on the raised earth behind him. The calabash rocks and releases its water into the small canal running to his vegetable plot. The positioning is always perfect and the rhythm hypnotic. However, Sabiou, one of the last remaining farmers still using the calabash, said he wished to own a treadle pump (below right).

Manzo Kanta Adam, a neighbour now in his 60s, explained that the land covered by immaculate vegetable plots used to comprise a small lake which diminished over the years and finally disappeared completely. As a child, Manzo and his neighbours would carry water from the lake for their sugar cane, tomato, and spice crops. As the water dried, the farmers followed it by digging wells, which were lined with wood and later cement. Treadle pumps were introduced several years ago. Manzo is one of many owners of treadle pumps in the area. He is proud of the fact that he has kept the pump maintained since he bought it, either by undertaking repairs himself or taking it to local artisans.

Danert (2006)

---

branching out to drill wells in people's yards, intended primarily for domestic use. In most cases these appear to be fitted with simple bailers. In cases where projects are supporting the introduction of domestic hand-drilled wells, they are being fitted with rope or hand pumps. These tend to be communal sources, and are subsidized, usually through development aid.

The subsidized sources are protected with concrete aprons, intended to prevent backflow of contaminated water down the well. On the other hand the privately financed wells do not have this protection, and the bailers risk being contaminated by being dropped on the ground. To a professional, the privately financed wells are poor quality, and would not be counted as 'improved'. To the users, they are providing accessible water—a key issue when alternative sources are situated a considerable distance from the home.

Figure 4.2 Hand auger drilling.

Small-scale irrigation through river diversion in Ethiopia

Dorward *et al.* (2004) point to the need for more intensive and drought-tolerant agriculture in Ethiopia, and it is highly favoured by Government (Carter and Danert 2006). Small-scale irrigation is a key approach to this, particularly given the failures of large-scale agriculture in sub-Saharan Africa in the 1970s. As in Niger, most farmers in Ethiopia rely on rain-fed agriculture. There are many cases of people irrigating their crops in Ethiopia, although the full extent has not been documented.

Professionals draw fairly clear distinctions between traditional and modern irrigation (Table 4.1). Interestingly, it is not the time that the technology has been used which provides the definition, but the infrastructure. Thus even if a group of farmers started to irrigate their crops as little as two years ago, but use stones and brushwood as structures, this will be referred to as traditional irrigation.

Figures 4.3 and 4.4 provide contrasting examples of what are considered to be modern and traditional river diversion structures. In the case of Kokono, the 'modern' scheme was initially constructed at a time of severe food shortage. The NGO managing the project was not concerned about the type of crops grown, as long as hunger could be alleviated. However, this, coupled with the sheer extent and type of construction, has resulted in dependency on the NGO for major repairs, such as masonry work and re-plastering. In contrast the farmers of Yandafaro are digging a 'traditional' channel

Figure 4.3 A diversion structure, Kokono, in north Ethiopia.

Figure 4.4 Construction of a spate intake, Yandafaro, in south-west Ethiopia.

to try and tap flashy, spate river flows for the irrigation of their crops. Unfortunately previous experience has been that the silt-laden, fast-flowing rivers have washed out such structures, leaving the fields high and dry above the watercourse (Carter and Danert 2006).

In cases such as Yandafaro, there is a yearning for the 'permanent' structures associated with 'modern' irrigation. However, as Kokono shows, not all 'modern' irrigation structures are permanent, particularly if the users cannot readily maintain them. There are also cases of modern schemes, or parts of schemes rendered totally dysfunctional due to massive damage from heavy river flows (Carter and Danert 2006). Some key stakeholders have indicated that only about 30 per cent of modern irrigation schemes in Ethiopia are actually still functioning, with failures resulting from seepage, siltation, drought, environmental hazards, labour shortages, poor administration, and the high speed of construction rather than the technology per se (Carter and Danert 2006).

Determining whether a modern or traditional irrigation scheme (or some mix of the two) is the most suitable is not simple. Unfortunately, the consultant is likely to be trained in modern methods, which can mean that small incremental improvements to existing irrigation schemes may be easily overlooked. Box 4.2 provides a contrasting story from Sero, where the NGO FARM-Africa took several years to learn from its mistakes.

Conclusions

The distinction between improved and unimproved domestic water supplies is a useful mechanism for considering the problem of inadequate access from a statistical point of view, and drawing attention to the plight of millions of people without a clean or close supply. However, when it comes to seeking ways to improve the situation on the ground, a much more nuanced approach needs to be taken. Ideally, this should consider each community on a case-by-case basis, and seek that they can be assisted to help themselves. The rush to try and meet the targets set in the United Nations Millenium Development Goals (MDGs), coupled with inadequate human and financial resources at the lowest levels of local government, means that this is not taking place to the extent that it could be. The Niger story points to the possibility of linking water users, local enterprises, and technologies in a way where all can benefit.

In the case of small-scale irrigation, the textbook distinction between modern and traditional schemes may serve the reporting requirements of projects and programmes, but misses out the possibility of supporting countless farmers who have taken the first steps towards watering their crops.

In the case of domestic and agricultural water supplies (as well as the area in between), incremental improvements, largely financed by water users themselves (through savings or loans) may provide some answers to the challenge of enabling people to access sustainable water supplies.

BOX 4.2 THE NGO FARM-AFRICA IMPROVES LIVES OF COMMUNITIES IN SERO, TIGRAY, NORTH ETHIOPIA

In the dry season, viewed from above, Sero valley stands out as a green oasis among the brown and bare fields (top right). According to local farmers and ex-FARM employees, the gully plugging raised the level of the gully considerably. One farmer, who worked as a technician on the project, claims that it was raised by 7 metres at one point (bottom right). Initial stone plugging of the gully was washed away in the first year, but subsequent stone gabions built in phases over a 3-year period succeeded in raising the level of the river. The gully plugging was undertaken in conjunction with the construction of a small-scale irrigation scheme and formation of women's goat groups, the provision of seeds for leguminous forage plants, and a nursery for seedlings.

When we visited Sero in May 2004 community members were fetching water from a spring within the gully to water in elephant grass along the bank further up the gully. Ato Bero, a local farmer, informed us that what used to be a big river now forms a lake in the rainy season, thanks to the trapping of soil by the gabions. Dense vegetation now grows in large sections of the river bed, which is used for grazing cattle, sheep, and goats.

People have been planting trees, and vegetables are cultivated with irrigation water, raised by a motorized pump from a pond in the gully. Farmers informed us that the pond could only be constructed thanks to the soil trapped by the gabions. Although the pump has not broken down over the three years they have operated it, we were informed that there are four people in the community who have been trained in pump maintenance, and that spares can be procured from Mekelle, the regional capital. The communal pump is used to water the fields of 42 farmers, who each contribute to fuel according to the size of their plot. The Sero farmers readily sell their produce (tomatoes, onion, green pepper, chilli, maize, oranges, and papaya) in nearby Enticcio.

Carter and Danert (2006)

■ SUMMARY

• Niger and Ethiopia are among the poorest countries of the world, on several different measures.

• Water supply technologies tend to be categorized as either improved or unimproved for domestic use, and traditional or modern for irrigation purposes.

• Case studies from Niger and Ethiopia show how this categorization can be an over-simplification that overlooks important issues of financing, maintaining, and managing sustainable water supplies.

■ REFERENCES

Carter, R. C. and Danert, K. (2006) *Small Scale Irrigation in Ethiopia: Experiences, Issues, Guidelines and Strategic Options for FARM-Africa*. Cranfield University at Silsoe, unpublished report.

Cotton, A. and Bartram, J. (2008) Sanitation: on or off track? *Waterlines* 27(1) Practical Action Publishing, England.

Danert, K. (2006) *A Brief History of Hand-drilled Wells in Niger: Only the Beginning*. Field Note published by the Rural Water Supply Network (RWSN)/Water and Sanitation Programme-Africa (WSP-AF).

Dorward. A., Kydd, J., Morrison, J. and Urey, I. (2004) A policy agenda for pro-poor agricultural growth. *World Development* 32(1): 73–89.

FAO (2008) *AQUASAT—FAOs Information system on water and agriculture*. United Nations Food and Agriculture Organization (FAO). Available from http://www.fao.org/nr/water/aquastat/ water use/index.stm, accessed 28 May 2008. Rome: United Nations Food and Agriculture Organisation (FAO).

Naugle, J. (1996) *Hand Augered Garden Wells*. Baltimore, MD: Lutheran World Relief.

UNDP (2006) *Human Development Report 2006. Beyond scarcity: Power, poverty and the global water crisis*. New York: United Nations Development Programme.

UNDP (2008) *Statistics of the Human Development Report*. New York. United Nations Development Programme. Available from: http://hdr.undp.org/en/statistics/, accessed 6 February 2009.

UNFPA (2008) *Population, Health and Socio-economic Indicators*. New York: United Nations Population Fund (UNFPA). Available from: http://www.unfpa.org/worldwide, accessed 28 May 2008.

UNICEF (2008) *Statistics by Area/Child survival and Health*. Available from http://www.childinfo. org/mortality_underfive.php, accessed 28 May 2008. New York: UNICEF.

UNICEF/WHO (2008) *Joint Monitoring Programme for Water Supply and Sanitation*. New York: UNICEF/World Health Organisation (WHO). Available from http://www.wssinfo.org/en/ 26_wat_leastDev.html, accessed 28 May 2008.

■ FURTHER READING

The Federal Democratic Republic of Ethiopia Special Country Programme Phase II (SCP II) Interim Evaluation, April 2005 Report No. 1643-ET. Can be downloaded from http://www.ifad. org/evaluation/public_html/eksyst/doc/country/pf/ethiopia.pdf

Carter, R. C. and Bevan, J. E. (2008) Groundwater development for poverty alleviation in sub-Saharan Africa. In Adelana, S.M. and MacDonald, A. M. (eds) *Applied Groundwater Studies in Africa*. International Association of Hydrogeologists Selected Papers.

Sutton, S. (2007) The risks of a technology-based indicator for the MDGs. Paper presented at the 33rd WEDC Conference in Accra, Ghana. Available from http://wedc.lboro.ac.uk/conferences/pdfs/33/Sutton_S.pdf

FAO (1989) NGO Casebook on *Small-scale Irrigation in Africa*. R. Carter (ed.). Rome: Food and Agriculture Organization.

■ USEFUL WEBSITES

For information on incremental improvements to domestic water supplies visit **http://www.rwsn.ch**. Click on Activities and then Self Supply where you will find a wide range of documents.

Visit the website of the Food and Agriculture Organization (FAO) **http://www.fao.org** and search for small-scale irrigation.

5

The souvenirs of communism: missed opportunities for sustainable development innovations in the enlarged European Union?

Petr Jehlička

Introduction

One autumn morning in the early 1990s a group of professionals from Slovakia and the Czech Republic arrived at the vast Open University (OU) campus in Milton Keynes, UK to be trained to teach the OU's environmental courses in their countries. As the bus drew up to the campus, their jaws dropped at the sight of car parks surrounding university buildings filled with some 2,000 cars. 'So we came *here* to learn how to protect the environment?' sighed one of the Slovak visitors.

This brief anecdote from an OU colleague provides several points I wish to draw on in developing the argument of this chapter, which concerns the former socialist countries of Central and Eastern Europe (CEE). First, from the start of the process of democratization of these countries in 1990, the assumption both in the West and CEE was that former socialist societies needed to adopt Western know-how, policy approaches, and models if they were to lift themselves out of the misery inflicted on them by the previous regime. The phrase 'transition to democracy and market economy' used both in academic writing and in assistance programmes was a strong evocation of the linearity of this process. In the context of eastern enlargement of the European Union (EU), 'catching up with the more advanced practices of their west European neighbours' (Baker, 2006: 207) was the dominant perspective.

This applied to all policy spheres but first and foremost to the environment. Both Western media and CEE environmental activists used powerful images of smoke stacks churning out dark clouds of smoke with ensuing adverse effects on human health and on ecosystems, such as tracts of dead forests killed by acid rain, as the ultimate evidence of the communist regime's failure to deliver on its promises for harmonious social development. As Zsuzsa Gille, a Hungarian-born US sociologist observed:

> the textual, visual and statistical representations all suggested, therefore, that state socialism was wasteful, both in the sense of squandering resources and in the sense of being full of wastes: producing too many rejects, too much waste and garbage, and too many outdated and superfluous goods.

> Gille (2007: 2–3)

Thus, explicitly or implicitly, in these representations, much of the environmental damage was attributed to the production side of the economy.

It is beyond question that some structural features of socialist states economies, such as reliance on low-grade coal as a source of energy generation, affected the quality of the air, water, and soil. However, some observers with firsthand experience became critical of these media representations and their political implications. Manser (1993: 18) referred to them as 'one of the last propaganda coups of the cold war' and Pavlínek and Pickles (2004: 242) described them as Western myths of ecocide, toxic nightmare, and ecological disaster in CEE. As Gille (2007: 3) pointed out, while dirt, toxic waste, wastefulness, and degraded nature in the East was contrasted with cleanliness, efficiency, and thriftiness of the West, some 400,000 tonnes of detritus and 40,000 tonnes of toxic waste per year were exported from West Germany to East Germany alone.

Thus the somewhat simplistic binary relationship between dirty and wasteful East and clean and efficient West begins to look less convincing. It will be my goal in this chapter to unsettle these conventional representations and offer an alternative and more complex view of the relationship between the West and CEE in this area.

The demise of the CEE preventative approach to waste management

As the introductory anecdote suggests, there were some social practices in CEE countries that, although not necessarily motivated by environmental concerns, were quite advanced from the sustainability point of view, but these rarely received attention in scholarly works or in policy exchanges and strategies. As Gille (2007: 3) pointed out, the recollections of her youth in socialist-times Hungary, such as waste collection campaigns in school, returning empty bottles to grocery shops, and taking one's own bags when doing shopping, did not square with the images of socialist societies as hopelessly wasteful which she encountered after her emigration to the United States in 1988.

In fact these practices, as well as the general approach to waste management in socialist Hungary, were strikingly similar to the declared waste management policies of the EU, which Hungary was supposed to adopt as part of its accession to the Union in 2004.

EU environmental management policies are officially guided by the 'waste management hierarchy' with environmental risks associated with these methods increasing from 1 to 5 (see Chapter 14):

1. Reduce—don't generate waste in the first place
2. Reuse—use products for their original design purpose more than once
3. Recycle—reprocess waste materials to manufacture new products
4. Recover—burn wastes to recover energy for heat and power generation
5. Dispose—landfill wastes that are not suitable for recovery, recycling or reuse.

The state socialist approach to waste management could be described as preventative because it was based on waste minimization, reuse and recycling. In terms of the EU waste management hierarchy, this was a highly progressive approach. Landfilling, while widely used, was being discouraged by the state more than in the post-socialist period and incineration was virtually non-existent: socialist Hungary had only one purpose-built waste incinerator. The recycling rate for hazardous waste in 1980s Hungary was 20 per cent while in the then 12 EU countries it was 8 per cent. The most environmentally risky method of management of hazardous waste—incineration—accounted only for about 3 per cent of its volume (Gille 2007: 186). In keeping with thriftiness practices in people's daily life as mentioned above, the volume of municipal waste per capita in Hungary was lower than in most EU member states at the time.

At this point you might feel compelled to ask to what extent these environmentally favourable features of waste management, which were compatible with the approach promoted by the EU, were considered and further developed during the process of Hungary's accession to the EU. The frustrating answer is: not at all. In fact, they were scrapped and replaced by a new approach based on the least environmentally friendly methods of waste management—incineration and landfilling. In reality, rather than seizing eastern enlargement as an opportunity to strengthen EU environmental strategies advocating prevention, reduction, and recycling, it was seen primarily as an opportunity for the expansion of the market for EU end-of-pipe technologies (i.e. treating pollutants at the end of the production process after they have been formed). There were two crucial prerequisites for achieving this conceptual U-turn. First, the branding of preventative socialist waste management policies domestically as backward and incompatible with European-ness and European consumerism (Gille 2007: 201). Second, the projection of the image of post-socialist Hungary within the EU as a country lacking any meaningful waste legislation, lacking technical infrastructure to implement modern waste management methods and being incapable of implementing any progressive change (Gille 2007: 188).

Sustainable consumption practices in CEE and EU sustainable consumption policy

The opportunity to use the CEE approach to waste management as a source of policy innovation in the whole EU, however theoretical this opportunity was, was lost long before the accession of CEE countries to the Union in 2004. Let me now consider another and more recent example of the interaction between the EU and CEE in the area of sustainability-compatible practices—sustainable consumption.

Since the establishment of sustainable consumption as part of the global environmental governance agenda at the 1992 United Nations Conference on Environment and Development (**UNCED**), it has been increasingly viewed as a prerequisite for the achievement of sustainable development (Fuchs and Lorek 2005). The solutions proposed by UNCED included promoting eco-efficiency and using economic instruments to shift consumption patterns, but it also called on governments to develop 'new concepts of wealth and prosperity which allow higher standards of living through changed lifestyles and are less dependent on the Earth's finite resources and more in harmony with the Earth's carrying capacity' (UNCED 1992: Chapter 4, paragraph 4.1).

Thus sustainable consumption was initially a concept that embraced two competing perspectives—one reformist, and one more radical implying realignment of social and economic institutions (Seyfang 2005). The former perspective, often referred to as green consumerism, is based on the premise that individual consumers' choices drive market transformation towards the provision of greener goods and services and thus greater resource efficiency. The problem with the reformist approach is that efficiency gains alone can be nullified and reversed by a growth in consumption volumes. This problem is addressed by the latter perspective which implies reduced consumption levels with attendant lifestyle changes. A simple illustration of the two perspectives is driving a car further with lower fuel consumption per 100 km versus travelling by train or not at all.

EU environmental policy of the 1970s and 1980s focused primarily on the regulation of production via emissions-limit values and on technology standards. Following the realization that this left a number of environmental problems unresolved (such as those arising from the growing use of private transport), the EU began to develop legislation addressing environmental problems from the perspective of consumption. The EU took part in wider international efforts which culminated in the World Summit on Sustainable Development (**WSSD**) in 2002. The Plan of Implementation, the most important outcome of the WSSD, called on governments to 'encourage and promote the development of a 10-year framework of programmes in support of regional and national initiatives to accelerate the shift towards sustainable consumption and production' (UN/WSSD 2002: Chapter III, Clause 14). The March 2003 European Council identified the development of the 10-year Framework Programme as one of the key priorities of the EU and described it as a primarily internal challenge: 'putting our own house in order by delivering at home what we would like others to do too' (European Commission 2004: 5).

Sustainable consumption-compliant food practices in the Czech Republic

Focusing on another post-socialist CEE country, the Czech Republic, I will briefly introduce two types of social practice compatible with sustainable consumption. Voluntary simplicity and food self-provisioning, while not universally practised, are familiar to large sections of the Czech society. The term 'voluntary simplicity' refers to the choice out of free will, rather than by being coerced by poverty, to limit expenditure on consumer goods and services, and to cultivate non-materialistic sources of satisfaction and meaning. Etzioni was unequivocal about voluntary simplicity's environmental benefits: 'There can be little doubt that voluntary simplicity, if constituted on a large scale, would significantly enhance society's ability to protect the environment' (Etzioni 1998: 638).

Voluntary simplicity in the Czech Republic was a subject of long-term research by the sociology professor Hana Librová. In 1992, she carried out in-depth interviews with people in 49 households whom she described as 'people who resist the general idea of a society oriented towards attaining high levels of consumption' (Librová 1994: 207). A decade later she interviewed 15 households selected from the original sample (Librová 2003). Most of the respondents were people who had moved, in some cases several years before the 1992 research commenced, from comfortable conditions in towns to live in the country and settled individually in villages or rural towns. The furnishing of their flats was markedly modest, even poor, and they chose not to have televisions. The respondents used bicycles rather than a car. The great majority of them had a secondary or university education but this was not used professionally. Many of them were vegetarians and relied on self-provisioning for their fruit and vegetables.

In comparison with voluntary simplicity, food self-provisioning is a more widespread social practice in the Czech Republic. While the exact scope of environmental benefits of food self-provisioning is difficult to establish, they certainly include a shorter distance from the garden to the table than for food acquired in a conventional way in shops, and often a virtually organic standard of cultivation (Jehlička and Smith 2007).

Self-provisioning on allotments or smallholdings outside the cities originated in earlier times, but expanded considerably in the socialist period after the Second World War. Fresh fruit and vegetables were expensive, quality was poor, and supply patchy. Self-provisioning in this period also worked at a political–cultural level to make space for an element of independence from state organization and provision of both food and work (Haukanes and Pine 2004: 108).

However, food self-provisioning did not disappear with the arrival of Western consumerism in the 1990s. Evidence from a national survey of 1,100 respondents conducted in February 2005 showed that this style of provisioning is still widely practised: 41.5 per cent of the population use a garden or allotment to produce vegetables and fruit for their own consumption (Smith and Jehlička 2007). In his study of productive gardening in Slovakia, Smith (2002) demonstrated that hardship was not the only motivation for self-provisioning. He showed that both during state socialism and in the years since, the non-economic reasons for these practices have been at least as prominent as the economic: 'household food production can only be understood in relation to the constellation of household, cultural/historical and economic (not only capitalist) forces' (Smith 2002: 244). Indeed the evidence from Smith and Jehlička's (2007) Czech household interviews and quantitative survey work confirm these Slovakian findings. Far from being directly related to austerity in the economy, the Czech evidence suggests that there are higher rates of self-provisioning in more financially secure households than not. The proportion of people with high living standards that grow their own food is higher (43.6 per cent) than the proportion of people doing so with the lowest living standard (35 per cent).

In Prague, the capital of the Czech Republic, 21 per cent of people grow their own fruit and vegetables. In mid-size towns this proportion is 41 per cent and in villages with less than 2,000 inhabitants, 65 per cent of people are involved in this kind of gardening. People's main motivation for self-provisioning is stated as being about having access to their own healthy food. The second motive is financial and the third is that it is a hobby. As a result, self-provisioning of many commodities is very high. For example, the 2005 national survey showed that, among people involved in self-provisioning, two-thirds of the consumption of blackcurrants, strawberries, and apples is accounted for by people's own production (Smith and Jehlička 2007).

The transfer of EU sustainable consumption policy to the Czech Republic

Sustainable consumption as a policy concept arrived in the Czech Republic via two inter-related initiatives in 2003, just a year before the country's accession to the EU—one promoted by the United Nations Environment Programme (UNEP) and the other driven by the EU. Following the publication of the 2001 UNEP strategic policy report *Consumption opportunities* and the emphasis placed by the 2002 World Summit on consumption, UNEP ran a series of seminars under the title SCOPE (Sustainable Consumption Opportunities for Europe) in selected European countries including, in May 2003, the Czech Republic.

UNEP's one-off initiative in the Czech Republic soon transformed into part of the EU 10-year Framework Programme. In anticipation of the country's upcoming EU membership, the Czech government set out to develop the 10-year Action Plan for Sustainable Consumption. After several years, this resulted in the Framework Programme of sustainable consumption and production of the Czech Republic, which was conceptually based on EU documents on sustainable consumption.

In contrast, a range of more hands-on activities followed more immediately after the 2003 seminar. Two pilot projects, small in scope but nonetheless logistically complex, were implemented in parallel. At the Ministry of the Environment in Prague and at the Ombudsman Office in Brno (second-largest city in the Czech Republic), some aspects of the operation were 'ecologized'. This included switching to recycled copy paper and fitting the buildings with energy-saving light bulbs. In the area of food, the internal directive stipulated that only drinks in returnable bottles and only organic food would be used at official functions organized by the Ministry (Vondrouš 2004; Kašpar 2004).

The greatest challenge proved to be the plan to introduce an organic meal option to the menu of the Ombudsman office's canteen. Developing the supply network of organic ingredients for the organic lunch option in a country with most of its 800 organic farms specializing in meat production for export turned out to be a difficult task. Subsequently, the organic meal option was also introduced in a Brno kindergarten. Nevertheless, despite this huge effort, after about six months both organic food schemes were dropped due to low demand for the organic meal option, mainly due to its substantially higher price (Kanichová 2004).

Thus, in terms of the two perspectives introduced earlier in this chapter, the projects and activities on sustainable consumption promoted in the country by UNEP and the EU were firmly of the first—reformist—conceptualization. Indeed, leading figures of the Czech sustainable consumption initiative appeared sceptical about reduced levels of consumption as a basis for policy (Smith and Jehlička 2007). Instead, when voicing their ideas on policy proposals for sustainable consumption, they invoked a vocabulary of neoliberalism, a vocabulary which includes promoting market competition, choice, citizen-consumerism, and individual responsibility of citizens for the implications of their consumption. For example:

> *a group of people is emerging in the Czech Republic to whom status is not demonstrated simply by a new house, but by a house built according to ecological principles. These are the people who are sufficiently well off to afford it, rather than the category of people with an alternative lifestyle who leave the city for the countryside seeking to be self-sufficient and live independently from the external world.*
>
> *Kanichová (2004)*

Conclusion

I hope that I have managed in these two brief case studies to show you that a more in-depth approach to investigating environmental policies and practices in former socialist countries of CEE reveals a much more complex situation than that depicted by western media and consultants. As the case of Europeanization of waste management in Hungary demonstrates, this depiction conveyed a *tabula rasa* (clean slate) approach (Gille 2007: 196) that assumed environmental policy was virtually non-existent in CEE countries. It nurtured the general approach of Western aid agencies and of the EU towards the environmental reform of CEE namely, the wholesale adoption of western solutions. The striking element of the waste management story is that there was already a policy in place in the CEE and it was highly compatible with the approach promoted by the EU in its programmatic documents. However, rather than seeing the CEE approach to waste management as supporting EU declarations, economic rationality that saw the CEE as a market opportunity for EU countries' industries—incineration and landfilling technology—prevailed.

As the Czech case of food self-provisioning shows, widespread CEE practices which correspond with strategic goals of the EU in the area of sustainable consumption are neglected both by EU institutions and CEE domestic policy actors. Instead, policy innovations of Western provenance with marginal sustainability potential are prioritized. The common practice of virtually organic food self-provisioning is disregarded by the policy community which focuses, instead, on market-based green consumerism. The fact that the preventative approach to waste management and food self-provisioning both defy the underlying principles of the prevailing market-driven economic paradigm perhaps offers a clue for their neglect as a source of policy inspiration.

■ SUMMARY

- In contradiction to the views widely held by the EU and other Western promoters of sustainable development, in some areas state socialist societies of CEE developed innovative policies and harboured social practices that were compatible with principles of sustainable development.

- In the area of waste, with respect to both production and consumption, the approach of CEE societies was based on the notions of prevention and thrift.

- In the area of food consumption, virtually organic self-provisioning is still widely practised.

- Rather than developing these domestic sustainability-compatible approaches and practices, CEE societies are expected to adopt Western sustainable development know-how, knowledge, and policy approaches in the name of Europeanization and modernization.

■ ACKNOWLEDGEMENT

I borrowed the phrase in the title from the book by Dubravka Ugrešić (2007) who, however, uses it in a different context and meaning.

■ REFERENCES

Baker, S. (2006) *Sustainable Development*. Abingdon, Routledge.

Etzioni, A. (1998) Voluntary simplicity: characterization, select psychological implications, and societal consequences. *Journal of Economic Psychology* 19(5): 619–643.

European Commission (2004) *Sustainable Consumption and Production in the European Union*. Luxembourg: Office for Official Publications of the European Communities.

Fuchs, D. A. and Lorek, S. (2005) Sustainable consumption governance: a history of promises and failures. *Journal of Consumer Policy* 28(3): 261–288.

Gille, Z. (2007) *From the Cult of Waste to the Trash Heap of History: The Politics of Waste in Socialist and Postsocialist Hungary*. Bloomington, IN: IUP.

Haukanes, H. and Pine, F. (2004) Ritual and everyday consumption practices in the Czech and Polish countryside: conceiving modernity through changing food regimes. *Anthropological Journal on European Culture* 12(1): 103–130.

Jehlička, P. and Smith, J. (2007) An unsustainable state: contrasting food practices and state policies in the Czech Republic. Paper presented at the IBG conference, London, 29–31 August.

Kanichová, K. (2004) Unpublished interview, Vita Foundation, Ostrava, 16 August.

Kašpar, J. (2004) Unpublished interview, Ministry of the Environment, Prague, 31 May.

Librová, H. (1994) *Pestří a zelení (Kapitoly o dobrovolné skromnosti)* (The colourful and the green: chapter on voluntary simplicity). Brno: Veronica and Hnutí Duha.

Librová, H. (2003) *Vlažní a váhaví (Kapitoly o ekologickém luxusu)* (The lukewarm and the hesitant: chapters on ecological luxury). Brno: Doplněk.

Manser, R. (1993) *Squandered Dividend. The Free Market and the Environment in Eastern Europe*. London, Earthscan.

Pavlínek, P. and Pickles, J. (2004) Environmental pasts/environmental futures in post-socialist Europe. *Environmental Politics* 13(1): 237–265.

Seyfang, G. (2005) Shopping for sustainability: can sustainable consumption promote ecological citizenship? *Environmental Politics* 14(2): 290–306.

Smith, A. (2002) Culture/economy and spaces of economic practice: positioning households on post-communism. *Transactions of the Institute of British Geographers* 27(2): 232–250.

Smith, J. and Jehlička, P. (2007) Stories around food, politics and change in Poland and the Czech Republic. *Transactions of the Institute of British Geographers* 32(3): 395–410.

Ugrešić, D. (2007) *Nobody's Home*. London: Telegram.

UN/WSSD (2002) *Plan of Implementation*. New York: United Nations.

UNCED (1992) *Agenda 21: The United Nations Program of Action from Rio UN Publications*. Available from: http://www.un.org/esa/sustdev/agenda21/english/agenda21toc.htm, accessed 24 May 2008.

Vondrouš, D. (2004) Unpublished interview, Ministry of the Environment, Prague, 31 May.

6

Sustainable transport systems: learning from Cuba

James Warren

Introduction

Cuba (Figure 6.1) offers a unique perspective as a case study of sustainable transport practices. Although the country has many characteristics similar to other developing nations, it is alone in the fact that it has endured one of the longest economic blockades in recent history. Thus, since 1960, Cuba has faced enormous pressures from its nearest potential trading partner (the United States) due to the blockade, which has caused innumerable problems in all sectors of the economy. The blockade attempts to halt the majority of goods imported and exported to and from Cuba by making it illegal, according to US law, for businesses, goods, and people to travel freely to that country.

This chapter examines the effect that the blockade and subsequent events, such as the collapse of the Former Soviet Union (FSU), have had on the transport sector. These events have led to a transformation in how goods and people are moved, not least because of a huge reduction in the amount of hard currency available to pay for fuel, vehicles, and spare parts (Enoch *et al.* 2004) but also due to consequential changes in society.

These societal changes have resulted in a number of innovative behavioural and techno-logical outcomes both in transport and in food production, biotechnology, literacy, medical training, and practice. This chapter describes, therefore, the sustainable transport practices which have stemmed not only from the situation in Cuba but also from the ingenuity of the Cubans. It closes with some more speculative ideas about how Cuba's transport sector might develop in the coming years and the challenges that this presents in terms of sustainability.

Cuba in brief

The history of Cuba is shown in Box 6.1, using key dates to indicate some of the major events. Cuba was 'settled' by the Spaniards in search of new pastures for cattle grazing, and for grow-ing tobacco and sugar shortly after its discovery by Columbus in 1492 (Suchlicki 2002). Sugar

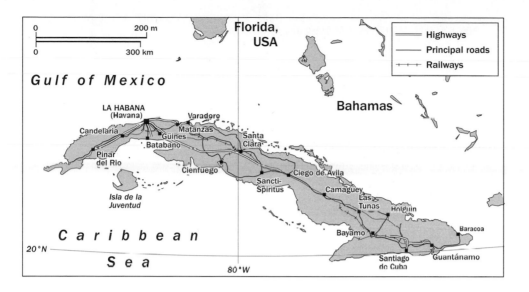

Figure 6.1 The Republic of Cuba. In general the road network is not as developed as many other countries with similar wealth levels, although the main highway running from the capital Havana, in the north-west, to Santiago de Cuba in the south-east, serves as the main conduit across the country. The railway also connects these two cities.

became a key industry through the use of enslaved labour for more than a century and it has been a key driver for technological change, including the introductions of the rail and road networks.

Table 6.1 summarizes some of Cuba's key demographics for the year 2006, with comparative data for 2004 for the United States and for Europe as an average of the 25 European community countries. The table shows that, in terms of road density, vehicles owned, and distance travelled per person, Cuba's figures are much lower than both the EU and the US. In addition, fuel is expensive per litre relative to the much lower overall income levels. Less fuel is used for the transport sector in Cuba as a percentage of total energy use. This is partly because Cuba has still not fully recovered from the withdrawal of the FSU. The country endured massive cuts of its imports during the restructuring of the FSU and many of these crashes in supply caused periods of extreme crisis (Alepuz and Valdés Ríos 2002).

The exact impact of the dramatic post-Soviet changes on travel patterns in Cuba is difficult to measure due to the large number of different responses, many of them informal, made both by Government and by individuals. However, if figures for the number of vehicles and the amount of oil used in the transport sector—perhaps the most obvious proxies for mobility— are examined, the effects were calamitous. Cuban imports as a whole fell in monetary value by 75 per cent, from 8.1bn Cuban Pesos in 1989 to 2.0bn Cuban Pesos in 1993, while over the same period in the transport sector, fuel imports were cut by a similar amount, and imports of transport equipment fell by 86 per cent (Farah 1992; Diaz-Briquets and Perez-Lopez 1995). The next section will focus on these general trends.

BOX 6.1 SELECTED KEY EVENTS IN THE HISTORY OF CUBA

Colonial period

1492 Columbus 'discovers' Cuba

1523 Slave trade begins with Africans arriving to work in the mines and fields

1595 Sugar mills are established by cattle ranchers

1607 Havana becomes the capital of Cuba

1886 Spain abolishes slavery

1898–1902 First automobiles arrive in Havana

1899 Spain relinquishes Cuba under the Treaty of Paris

Republican period

1902 The Republic of Cuba is proclaimed and US intervention begins to decline with the end of military occupation

1920 Collapse of the sugar boom

1952 General Batista seizes power through a military coup

1953 Fidel Castro attempts failed attack at Moncada

1958 US withdraws support for Batista

Revolutionary period

1959 Fidel Castro assumes command

1960 Cuba and Soviet Union re-establish diplomatic relations

1961 US breaks off relations with Cuba followed by a defeated invasion at the Bay of Pigs

1962 US formally declares embargo on trade with Cuba—the blockade begins in earnest

1962 The Cuban missile crisis causes a 'nuclear near miss' for the US and Soviet Union

1965 The Cuban Communist Party is created and Che Guevara initiates series of trips abroad

1967 Che Guevara is killed in Bolivia

1970 Sugar harvest is very poor causing productivity to fall tremendously

1977 US President Carter eases travel restrictions for US citizens going to Cuba, and sets up interest sections in Washington and Havana through various third-party embassies

1980 Cuba and Soviet Union enter bilateral cooperation agreement for 1981–1985 with increases in trade worth approximately US$8 billion per year

1988 Soviet Union cuts trade by some 12 per cent in the first quarter of the year; Cuba's economy shrinks by some 3.5 per cent in the entire year—the USSR begins to trim its aid budget to Cuba

1991 The Soviet Union falls and aid to Cuba stops suddenly

1992 The Special Period is announced focusing on a combination of austerity measures and pushing forward with both tourism and biotechnology markets to ease the economy

1992 The US tightens the embargo, limiting investment by third party countries linked to the United States—the blockade is written into US law

1993 Limited enterprises and more open market experiments begin to take place in some sectors but shortages in other industrial sectors cause severe hardship

1996 Further sanctions are placed on Cuba by the United States

2006 (July) Raul Castro assumes power under a temporary measure as Fidel undergoes major surgery

BOX 6.1 CONTINUED

2006 Energy savings and efficiency revolution is launched in Cuba
2008 Fidel Castro begins to step down as commander in chief and president of Cuba as Raul Castro formally takes over the presidential role
2019 Havana will celebrate the 500th anniversary of its founding

Adapted from Segre *et al.* (1997); Suchlicki (2002) and BBC (2008)

Table 6.1 Key demographics and typical transport indicators.

Demographic	Units	Cuba	EU 25	United States
Population (2006)	People	11,267,900	463,646,244	293,192,000
Size of economy measured as GDP per capita	US$/person	4,051	33,347	36,275
Area	km²	110,860	4,104,844	9,852,684
Population density	People/km²	102	176	32
Population growth rate	Percentage for 2000–2006	0.8	2.5	5.1
Urban population	Percentage of total population	75.5	72.8	80.5
Energy used for transport	Percentage of total energy consumed	~ 11	30.7	27.8
Fuel price	US$/litre	1.01	1.34	0.66
Road density	km of roads/km² area	0.51	1.80	0.70
Car ownership level	Total cars and light trucks per 1000 people	~ 34	470	777
Average annual motorized mobility	km/person/yr	~ 1,800	12,800	28,300

Sources: Metschies *et al.* (2007); ONE (2005); Eurostat (2007a, b); Davis *et al.* (2008); PRB (2007); IRF (2007); US Census (2008).

Transport and mobility during the Special Period in Cuba

The 'Special Period' was declared in 1992 at a time when the economy was in a state of free fall. The declaration by the Cuban Congress called on the support and action of the Cuban population in this 'Special Period in time of peace' as if they were at war. As one author described it: 'Havana looked like a war-torn city, full of fears, needs and frustrations' (Jatar-Hausman 1999: 41). Constant blackouts, long waits for public transport, and erratic water supplies were part of many peoples' everyday lives.

Since oil is the main source of transport energy, one useful way of understanding the Cuban situation is by looking at the total oil supplied (e.g. total petrol and diesel consumed) in Cuba before, during, and after the Special Period. Figure 6.2 shows that oil consumption peaked in 1989 and then fell by 80 per cent to a low in 1994. Public transport passenger trip numbers peaked slightly earlier in 1986 showing that oil consumption had some lag time associated with it, before the sharp drop, possibly due to the island's own supplies acting as a buffer. When these trip numbers are considered on a per person basis one finds that Cubans readily gave up their trips, probably those associated with

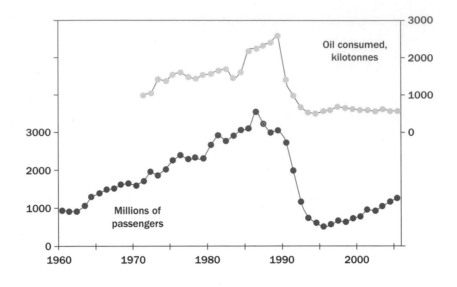

Figure 6.2 Millions of passengers transported each year on public transport (dark blue circles, left axis) and kilotonnes of oil consumed (light blue circles, right axis) from 1960 to the present.
Source: (IEA 2007; ONE 2005: Table XII.I Passengers transported by specialised state enterprises 2000–2005).

leisure activities, or switched from motorized transport to more sustainable modes such as walking and cycling, or reverting to animal traction. It is sometimes hard to imagine having to cut your travel by 80 per cent. For Cubans this meant changing from nearly a trip per day to one trip per week. These relatively low values of Cuban mobility are in sharp contrast to those observed in Europe, the US (see Table 6.1, final row), or even Malta with 3,875 km/yr by passenger car (Eurostat 2007a), despite its island boundaries. UK personal mobility was nearly 12,000 km/yr in 2006 (TSGB 2008) equating to some 2.8 trips per day.

In addition to scarcity of fuel, the absence of imported vehicles along with lack of parts, lubricants, and tyres meant that many buses simply did not operate. Some broken-down vehicles were dismantled in order to obtain parts for others. Things are now slowly beginning to change, as the bus sector has seen growth both in passenger numbers and in vehicle stocks (Mayoral 2006). New buses have been put into place on the principal routes in Havana city and this has also helped reduce waiting times and lower emissions overall.

Cuba kept moving

Cuba put a strong set of government-backed measures into place during the Special Period to ensure some mobility was guaranteed to all and to keep the economy going. As shown in Figure 6.2, these policies seemed to have worked: the economy has grown and mobility has also increased since the crisis period. In summary these resulted in (Enoch *et al.* 2004; Jatar-Hausman 1999):

- increasing occupancy in all motorized modes (Figure 6.3),
- replacing vehicle use with animal traction, bicycle, or walking modes, such as substituting oxen and mules for tractors, bicycles for buses and cars,
- reorganization of some public transport routes to support majority routes,
- creating more collective modes, such as shared taxis and use of organized, formal hitch-hiking methods,
- some employers began a system of worker routes using trucks and buses to collect their employees,
- reorganization of some job locations/homes to reduce travel distances,
- new working patterns such as working longer days over a shorter week to reduce their trips,
- complete restructuring of freight systems in order to reduce loads on the road and rail systems, and
- use of toll roads and taxis in tourist areas to generate capital for re-investment into public transport systems.

There has been concerted effort by government and transport operators to ensure that transport is accessible and affordable for all. Ticket prices for buses and shared collective taxis have remained very low with high occupancies of both cars and buses. This can

Figure 6.3 The Camello bus: one example of Cuban ingenuity applied to public transport. These buses are built in Cuba from recycled freight-hauling flatbed lorries. Camellos have a steel body fabricated onto the flatbed trailer which is then pulled by the cab (also called the tractor, or truck). The steel-bodied coach has two humps, hence the nickname camel buses. Although the buses carry a very large number of passengers, they have relatively high emissions. The camel buses were always meant to be a temporary solution and are being steadily replaced with new articulated buses.

result in long waiting times, but the populace understands, and they recognize that more equipment is required to meet everyone's travel needs. Outcomes for many Cubans included long waiting times, long commuting times, and very high modal shares of walking (>40 per cent in 1998) and cycling (>12 per cent in 1998) (Alepuz and Valdés Ríos 2002). For comparison, as an example, UK figures are < 3 per cent for walking and approximately 0.6 per cent for cycling (TSGB, 2008). Clearly human-powered modes are more sustainable and perhaps preferred over motorized modes for certain trips but they incur costs as well, such as accidents to cyclists and pedestrians, which must also be considered.

Lessons from Cuba for all

Using Banister's (Professor of Transport, Oxford University) framework of seven principles of sustainable transport (Banister 2005: 17–18) one can consider the measures Cuba has undertaken and try to map which principles are largely met by the policies undertaken. The seven principles can be summarized as:

1. reduce the need to travel,
2. reduce absolute levels of car use and road freight in urban areas,
3. promote energy-efficient modes for both road freight and road passengers,
4. reduce noise and emissions from vehicles,
5. encourage more efficient and eco-sensitive use of vehicles,
6. improve safety of pedestrians and all road users, and
7. improve quality of life for city dwellers.

Although these principles were devised for countries with higher car ownership, it is clear that they can apply to countries at different stages of motorization. It is not possible to correlate every policy to each of the principles, but there are some examples where the policies enacted in Cuba match these key tenets.

For example, the work done towards promoting energy efficiency (3, above) is probably seen within every one of Cuba's transport policies with an overall framework of reduced oil imports decreasing overall levels of use (see 2 above) and increasing vehicle occupancy in all modes of transport. Conversely, due to the blockade, the age of vehicles in Cuba is considerable and these antique vehicles have very poor fuel consumption compared to new vehicles in Europe. To some extent the Cuban transport authorities have also been striving hard to replace all the older buses with newer, more efficient articulated ones which have much lower emissions overall (see 5 above) but a full fleet replacement is not yet complete. There is much still to be done to continue the sustainable transport work started in Cuba stemming from the Special Period. Within some measures (6 and 7 above) bicycle lanes have been installed and some raised pavements have been refurbished. One of the major savings in transport energy has been from the decisions of Cubans to forego certain trips (measure 1, above) as revealed by surveys, in which many people cited trips that they had wanted to make but had not due to the Special Period conditions (Alepuz and Valdés Ríos 2002).

Table 6.1 showed that the US has nearly 1 car for every person, whilst Europe has nearly 1 car for every two persons. Cuba has about 1 car for every 29 people, so in that sense a car, or access to a car with fuel, is a highly valued commodity. In the 1950s, however, Cuba had the same levels of car ownership as the United States but post-revolution there was a marked stagnation. The United States is now considered saturated in terms of ownership (Dargay and Gately 1999), meaning that car ownership should begin to stabilize. This is an important point as it has been shown that increasing car ownership leads to increased traffic and mobility but also creates many other problems (e.g. accidents, pollution, etc.).

The level of car ownership in Cuba is interesting because one can speculate how it might develop in the future. If Cuba has a very high (saturated) level of ownership with almost everyone owning their own car it may result in much higher mobility in the short term, but eventually will result in congestion (WBCSD 2004). If Cuba remains with a relatively low vehicle ownership with high levels of bicycle, bus, walking, and shared modes of transport, the entire transport system will be much more efficient overall and should result in less congestion and fewer unwanted side effects such as pollution and accidents. Cuba acts as a showcase which breaks the typical cycle of transport growth based mostly on private car ownership.

Try to imagine your own city with its fuel or energy supplies cut by some 70 per cent—what might happen? In the UK there have been several fuel crises to date, partly due to refinery blockades by protesting drivers in industrial disputes. There has been concern about rising fuel costs and the cost of motoring in general. It would seem that many parts of Europe are stuck in a paradigm of automobile-based mobility sometimes called automobility (Urry 2004) in which the car is still regarded as central to personal mobility. Although the US blockade against Cuba has brought extreme hardship for so many, for so long, it has also allowed one very special and sustainable route of growth for Cuba. Cubans will have to ask themselves as the economy begins to grow: do they want to become like the US and many other countries that are so reliant on privately owned vehicles? Or will they go another way, with continued low motorization rates and low fares for public transport, following a more sustainable path?

■ **SUMMARY**

- Cuba has endured a huge decrease in transport investment and oil imports due to a long-standing economic blockade and the collapse of the Former Soviet Union.

- Cuba has restructured its transport system in order to provide the basic mobility requirements for all citizens.

- The islanders have had to invent novel ways of achieving mobility by creating a new mass transit system (camel buses) and formalizing car-sharing.

- By shifting to higher vehicle occupancies, using more public transport, and encouraging more walking and cycling, Cuba exemplifies sustainable transport practices.

■ ACKNOWLEDGEMENT

My thanks go to A. J. Lloyd for cartographic services.

■ REFERENCES

Alepuz, M. and Valdés Ríos, H. (2002) The urban transportation in time of crisis: the case of Havana. In X. Godard and I. Fatonzoun (eds), *Urban Mobility for All. Proceedings of the International Conference CODATU X*, pp. 547–551. Lome, Togo, 12–15 November, Leiden, Netherlands: CRC Press/Balkema Publishers, Taylor & Francis.

Banister, D. (2005) *Unsustainable Transport: City Transport in the New Century*. London: Routledge.

BBC (2008) Timeline: US-Cuba relations. Available from http://news.bbc.co.uk/1/hi/world/americas/3182150.stm, accessed 16 June 2008.

Dargay, J. and Gately, D. (1999) Income's effect on car and vehicle ownership, worldwide: 1960–2015. *Transportation Research Part A: Policy and Practice* 33: 101–138.

Davis, S. C., Diegel, S. W. and Boundy, R. G. (2008) *Transportation Energy Data Book (ORNL-6981)*, 27th edn. Knoxville, TN: Oak Ridge National Laboratories, and Washington DC, WA: Office of US Energy Efficiency and Renewable Energy, US Department of Energy. Available from http://cta.ornal/gov/data/tedb27/Edition27_full_Doc.pdf

Diaz-Briquets, S. and Perez-Lopez, J. F. (1995) The Special Period and the environment. *Proceedings of the 5th Annual Meeting of the Association for the Study of the Cuban Economy, Miami, Florida, 10–12 August. Cuba in Transition, volume 5*, pp. 281–292. Washington, DC: ASCE.

Enoch, M. P., Warren, J. P., Valdés Ríos, H. and Menoyo, H. (2004) The effect of economic restrictions on transport practices in Cuba. *Transport Policy* 11: 67–76.

Eurostat (2007a) *European Commission (2007) Panorama of Transport*. Luxembourg: Office for Official Publications of the European Communities.

Eurostat (2007b) *Population Estimates*. Available from http://epp.eurostat.ec.europa.eu/tgm/table.do?tab=tableandinit=1andplugin=1andlanguage=enandpcode=tps00001, accessed 20 October 2008.

Farah, D. (1992) Cubans are feeling unempowered. *The Washington Post* 21 December.

IEA (2007) *International Energy Agency data and the specific report World Energy Statistics and Balances*. Available from http://esds.mcc.ac.uk/wds_iea/, accessed 16 June 2008. Source of fuel consumption, Figure 6.2.

IRF (2007) *International Road Federation, World Road Statistics 2007, (Data 2000–2005)*. Geneva: International Road Federation.

Jatar-Hausman, A. J. (1999) *The Cuban Way: Capitalism, Communism and Confrontation*. West Hartford, CN: Kumarian Press.

Metschies, G. P., Friedrich, A., Heinen, F. and Thielmann, S. (2007) *International Fuel Prices 2007*, 5th edn. Available from http://www2.gtz.de/publikationen/isissearch/publikationen/details.aspx?RecID=BIB-GTZ071370, accessed 16 June 2008.

Mayoral, M. J. (2006) Cuba to buy more vehicles from China. *Granma International Edition*, 17 February 2006, available from http://www.granma.cu/ingles/2006/febrero/vier17/9yutong-i.html, accessed 28 Feb 2008.

ONE (2005) *Oficina Nacional de Estasticas, Cuba en Cifras/Cuba in Figures.* Havana: Oficina Nacional de Estasticas (ONE). Available from http://www.one.cu/publicaciones/cubaencifras/ccifras2005.pdf, accessed 16 June 2008.

ONE (2006) *Anuario Estidasticas de Cuba 2005, Edicion 2006.* Havana: Oficina Nacional de Estasticas (ONE). Available from http://www.one.cu/aec2006.htm.

PRB (2007) *Population Reference Bureau, World population data sheet 2007.* Washington, DC: Population Reference Bureau. Available from http://www.prb.org/pdf07/07WPDS_Eng.pdf, accessed 20 October 2008.

Segre, R., Coyula, M. and Scarpaci, J. L. (1997) *Havana: The Two Faces of the Antillean Metropolis.* Chichester, UK: John Wiley and Sons.

Suchlicki, J. (2002) *Cuba, 5th edn, from Columbus to Castro and Beyond.* Dulles, VA: Brassey's Inc.

TSGB (2008) *Transport Statistics Great Britain 34th edn.* London: Department for Transport.

Urry, J. (2004) The system of automobility. *Theory, Culture and Society* 21: 25–39.

US Census (2008) *USA Statistics in Brief—Population, Population Density and Area.* Available from http://www.census.gov/compendia/statab/cats/population.html, accessed 20 October 2008.

WBCSD (2004) *World Business Council for Sustainable Development, Mobility 2030: Meeting the Challenges to Sustainability.* Geneva: World Business Council for Sustainable Development.

▦ FURTHER READING

T. A. Henken (2008) Cuba, A *Global Studies Handbook*, Santa Barbara, CA: ABC-CLIO, Inc. This book offers anybody a thorough grounding in Cuba's geography, history, and institutions. It covers economics and development along with culture, music, and the arts. There is also an excellent bibliography and full timeline of events.

My Life (by Fidel Castro and Ignacio Ramonet) offers an unique insight to the thinking behind the Revolution (London: Penguin, 2008).

For students more interested in sustainable transport I recommend two new books which are both very accessible and interesting. D Metz (2008) *The Limits to Travel: How Far Will You Go?*, London: Earthscan Ltd, describes a very interesting UK based history of total trips, total mobility, and total travel time that each of us has inherently in our lives. Metz also looks at difficult issues about quality and quantity of travel using useful examples.

Lynn Sloman's *Car Sick: Solutions for Our Car-addicted Culture* (Green Books, 2006) offers plain-speaking sensible advice from a practitioner who knows both the dilemmas of travel and transport as well as some of the startling facts which make our transport lives so miserable.

▦ USEFUL WEBSITES

The Lexington Institute **(http://lexingtoninstitute.org/cuba_research.shtml)** publishes a wide variety of Cuba related reports; this link **http://lexington.server278.com/docs/cuba4.pdf** is to a 24-page report published in May 2002 called ' Survival story: Cuba's economy in the post–Soviet decade' (Philip Peters) which gives a good summary with photos for the period discussed in the chapter.

http://globalpublicmedia.com/the_power_of_community_how_cuba_survived_peak_oil accessed 12 Jan 2009. This is a short article by Megan Quinn (a permaculture agriculturist) who describes how many Cubans have used the collective agricultural campaign to increase food production, reduce fuel and pesticide use, and provide work. See also **http://www.powerofcommunity.org/cm/index.php**.

For students who enjoy learning through film, I recommend *The Power of Community: How Cuba Survived Peak Oil*, a short film (53 minutes, 2006, directed by Faith Morgan) which stems from the Institute for Community Solutions, a non-profit organization based in Ohio in the USA that designs and teaches low-energy solutions. The film describes itself as 'telling the story of the Cuban people's hardship, ingenuity and triumph over sudden adversity, through cooperation, conservation and community'.

7

At the coal face in Australia: the youth climate movement

Nicky Ison

Introduction

On the morning of 4 September 2007, 40 youth climate activists prepared to shut down the world's largest coal export port in Newcastle, Australia. Feeling nervous but exhilarated, wearing bright orange T-shirts that read 'We are the solar generation; You can't dig your way out of climate change', we organized ourselves, as planned, into three teams. Team one entered the coal port and dropped a large banner from one of the coal loading machines (Figure 7.1).

Meanwhile, the nine people in team two locked themselves onto other pieces of coal loading machinery. Outside the fence the remaining team (including me) formed the public face of the action, talking with local residents and being interviewed by local, national, and international media outlets. All told this action:

- stopped work at the Carrington Coal Terminal for four hours, preventing the export of 10,000 tonnes of coal;
- gained international media attention;
- inspired other young people to take direct action against the expanding coal industry; and
- brought into stark relief the inadequacy of the climate policies about to be discussed by world leaders three hours down the road in Sydney at the main meeting of the Asia Pacific Economic Co-operation (APEC) Forum.

These young people were not professional activists. Most had never before participated in a non-violent direct action. We were there because, while climate change is arguably the biggest threat facing the world today, international leaders and institutes such as APEC continue to expand the very industries that are causing the climate crisis. Our action was just one of hundreds being taken by people, young and old, to stop the expansion of the climate-polluting coal industry and just one of many things that the youth climate movement is doing to ensure a safe climate future. (For more examples of direct action against coal see Sourcewatch (2008).)

Figure 7.1 Banner drop, Newcastle Coal Port Action (Ramachandran 2007).

In this chapter, I propose that working towards a more environmentally sustainable world requires a diverse social movement, which must include a range of change mechanisms from lobbying bureaucrats to establishing renewable energy projects to grassroots direct action. Further, I show why it is essential to involve today's youth in such a movement. A case study of the Australian Student Environment Network (ASEN) is explored as a model for effective youth involvement and the utilization of diverse tactics in action. Specifically, I look at ASEN's climate change-related activism. I also establish the political context in which ASEN works. Please note, in this chapter I will not explain the threat that climate change poses to sustainability, development, and the environment; rather I will focus on what people can and are doing about it. I speak from personal experience as the National Convenor of ASEN in 2007.

Social movements

There is a huge body of literature that describes and demonstrates, through tangible examples, that social movements are integral to the prevention of environmental degradation and the development of a more sustainable society. For example, Diesendorf (2007: 332) states that movement strategies are essential to facilitate the social change required to achieve a safe climate and subsequently 'a better society and environment'. But what are social movements? Social movements have been defined as 'an organized effort by a significant number of people to change or resist change in some major aspects of society' (Scott and Marshall 2009) and as:

> collective actions in which the populace is alerted, educated and mobilized, sometimes over years and decades to challenge the power holders and the whole society to redress social problems and restore critical social values . . . which uses various different strategies over the course of its life.

Moyer et al. (2001: 2)

Australia provides a useful case study to demonstrate why such a movement is essential.

Australia's role and response to climate change

Some statistics:

- Australia has the highest per capita emissions in the world at 27.2 tonnes of carbon dioxide equivalent (t CO_2–e) per person compared with the United States at 21.4 t CO_2–e (Turton 2004) and China at 3.8 t CO_2–e (UNDP 2007).
- It is the largest coal-exporting nation, exporting 233 million tonnes of coal per annum, 30 per cent of global coal exports (Australian Coal Association 2007).
- The stationary energy sector, which covers 'the generation of electricity and the direct combustion of fuels for purposes other than transport' (Department of Climate Change 2007), accounts for 50 per cent of Australia's domestic emissions (Australian Greenhouse Office 2007).
- Since 1990 emissions in the stationary energy sector have risen by 42.6 per cent (Australian Greenhouse Office 2007).

These statistics indicate that Australia is an important contributor to climate change and that thus far its response has been insufficient to stop or even slow the rise in domestic greenhouse emissions. The economic reality behind these statistics is that unlike most 'rich world' nations Australia is currently highly dependent on resource-extracting industries. Further, since the 1980s, neoliberalism and its doctrine of the supremacy of market competition has become hegemonic at all levels of Australian governments and in both major parties. This hegemony has come at the expense of environmental protection and coincided in the last decade with a conservative Federal Government led by Prime Minister Howard. This government was characterized by climate scepticism, attacks on civil society, and a close relationship with polluting businesses. As such, for ten years major polluting industries effectively wrote Australia's climate change and energy policy (Pearse 2007; Hamilton 2007).

When governments refuse to institute policies to combat climate change because they are closely tied to businesses that directly profit from creating climate change, it requires a strong movement of people to step up and force the necessary changes.

The role of young people

Students and young people have an important role to play in the movement to avert anthropogenic climate change. As young people, we are often (though not always) less encumbered by commitments such as mortgages and a family and thus are able to take more risks both personally and politically. We have the passion and energy of youth. Our motivation is personal; it is my generation and my children's generation that will have to bear the consequences of today's inaction on climate change. Young people and students have a proud history to live up to; students have led movements that overthrew dictators, stopped the Vietnam War, slowed the proliferation of the nuclear industry, and more. The threat of climate change has the potential to eclipse all past victories, but students and young people through networks and organizations such as ASEN are rising to the challenge.

The Australian Student Environment Network

The Australian Student Environment Network (ASEN) is a decentralized network of environment collectives at universities, technical colleges, and high schools across Australia (ASEN 2006). ASEN in its current incarnation developed after an attack by the government on students organizing. In 2005, Prime Minister John Howard introduced a piece of legislation called Voluntary Student Unionism (VSU) or Anti-Student Organising Legislation. This legislation systematically removed funding for student organizations, out of which the student environment movement and environment collectives were run.

As a response to VSU, from being a loose network of collectives, ASEN became an incorporated, not-for-profit network, with independent funding (see below) and a paid national convenor. Subsequently, while other student movements have dwindled in the shadow of VSU, the student environment movement through ASEN grows in strength.

Mechanisms of change

At all of ASEN's training events there is one workshop that we always run—Theories of Change. This workshop is based on the premise that everybody has their own valuable experience and world view and thus everybody has a different theory about how they think change is made, whether they articulate it or not. The workshop is the start of a process for people to articulate their personal theory of change. It has been developed using a framework from the *Resource manual for a living revolution* (Coover *et al.* 1977), part of which asks people to consider which mechanisms for creating change they think are successful. Of these mechanisms, ASEN successfully utilizes tactics that fall under the following categories.

- Building alternatives
- Education
- Influencing decision makers to make better decisions
- Confronting power holders and making them give us what we want.

The next section details some examples of how these mechanisms are practised through ASEN's climate activism.

Complementary approaches

There are two main aspects to ASEN's climate activism—capacity building and campaigning. Both are essential to creating change. Capacity building includes:

- creating the organizational structures and resources that support student organizers, facilitate decision making and enable coordination and collaboration across Australia; and
- education and training programmes that ensure new and existing organizers have the tools and networks they need to create change.

Within ASEN we undertake this organizational development work in a way that fits with the building alternatives and education mechanisms of creating change.

ASEN's two major climate campaigns are the Clean Energy on Campus Campaign and the No New Coal Campaign. The former campaign primarily utilizes the mechanisms of influencing decision makers to make better decisions, while the latter fits more closely with the confronting power holders mechanism. While employing different tactics both campaigns are complementary—one creates the change that is necessary, the other resists the wrong form of change. Further, ASEN's campaigning work would not be possible without the capacity-building tactics employed.

Building alternatives

ASEN as an institution attempts to be an alternative to existing organizational structures. We attempt to 'be the change we want to see' (Gandhi). The idea of being an alternative organization requires that ASEN ascribes to a theory of change which may be loosely articulated but is codified in our Charter and Constitution (ASEN 2006). The ways in which we are building an alternative include:

- Having network structures and processes that attempt to challenge institutional arrangements like hierarchies that we believe perpetuate environmental degradation. Instead ASEN has a flat grassroots structure; each collective and ASEN working group recognizes the autonomy of the collectives and working groups involved and in return each collective is accountable to the ASEN charter.
- Employing participatory or direct democracy processes, including consensus decision making and a delegate- (not representative-) based, national council.
- Undertaking campaigns and initiatives that are decided on and driven from the grassroots, based on the interest and passion of the members and what is locally relevant.
- Being wholly run by students and young people, demonstrating that collective self-determination is possible.

Trying to create alternative structures that empower young people and create change is not always easy. A common tension that arises is that, while creating a structure that is flexible and workable, ASEN also tries to fit within the legal system as an incorporated organization. We chose to incorporate as this enabled ASEN to become financially sustainable and consequently we are able to pay convenors who facilitate the functioning of ASEN. Our funding is derived from three main sources—a regular giving programme (Friends of ASEN), membership dues, and grants from philanthropic organizations. Nonetheless, we are committed to the structure we have—it generally works and challenges those involved to believe that a different more equitable and just way of organizing society is possible.

Education

ASEN has a commitment to education, providing spaces where young people can learn essential social change skills through practice and workshops. ASEN's major education initiative is the annual Students of Sustainability Conference (SoS), which has been running since 1991. SoS is the largest environment and social justice conference in Australia, bringing together between 600–800 students, community activists, indigenous people, academics, and more. SoS includes workshops, forums, campaign planning sessions, and meeting people in the dining area to tell stories and have debates long into the night. Informal learning spaces such as the SoS dining area are an important component of a successful environment movement. These spaces inspire, build relationships of trust, and rejuvenate people who have been struggling to create change.

Other events in the ASEN training programme include a week-long summer training camp that brings together over 70 new campus organizers from across Australia, twice-yearly state-based weekend skillshares, and high school climate conferences. The facilitators or educators at these events are predominantly members of ASEN: thus young people experience peer-to-peer learning.

ASEN recognizes that education is essential to creating social change; however, the model of education that ASEN practises is 'activist' or 'popular education' as distinct from mainstream educational practice. Activist education draws on the pedagogies of Paulo Freire, the Highlander School, and more (see Box 7.1).

Education in ASEN is closely linked with empowerment and action. It goes hand in hand with the other mechanisms of change utilized. For example, at Skillshare events, we often role-play actions, where the participants practise the theory of direct action they have just learnt, and then have a space to reflect on both the theory and action.

Influencing decision makers to make better decisions

Through Campus Clean Energy campaigns ASEN aims to influence or lobby university decision makers to reduce university greenhouse emissions. Tactics employed in these campaigns include petitions, student referenda, forums, and lobbying meetings and creative stunts such as a dance competition between super solar (someone dressed in a sun costume) and cranky coal (someone dressed in a coal costume). These campaigns have been crucial to engaging thousands of young people. They have also been successful. For example, in October 2005, Monash University (based in Melbourne) Vice-Chancellor Richard Larkins agreed to meet the demands of a student campaign by creating a clean energy fund, committing Monash University to a 20 per cent reduction in energy use, and the creation of a staff position to oversee climate change and greenhouse reduction projects at the University (Monash University 2005; Monash Climate Change Group 2005).

BOX 7.1 PAULO FREIRE; THE HIGHLANDER SCHOOL

Born in 1921, Paulo Freire was a Brazilian adult educator who, among his many accomplishments, worked with peasant communities across Brazil on literary education. In his work *Pedagogy of the Oppressed* (1970) Freire challenges what he calls the banking model of education which he describes as 'an act of depositing, in which the students are the depositories and the teacher is the depositor. Instead of communicating, the teacher issues communiqués and makes deposits which the students patiently receive, memorize, and repeat'. He proposes that this oppressive model of education encourages passivity and does not encourage critical reflection. Instead, Freire proposed that education should be liberatory, based in dialogue, and socially transformative.

Since 1932, the Highlander School in Tennessee, USA, has worked with communities in the deep south, USA, struggling to overcome oppression. It was particularly instrumental in the civil rights movement bringing together white and black people to collaborate. One of Highlander's most famous students was Rosa Parks, catalyst for the Montgomery Bus Boycott. 'The Highlander school is founded on the principle that the answers to the problems facing society lie in the experiences of ordinary people. Those experiences, so often belittled and denigrated in our society, are the keys to grassroots power' (Highlander Center).

Confronting power holders and making them give us what we want

The ASEN coal campaign is part of a broader movement to stop the expansion of the coal industry in Australia and to build pressure to phase coal out. In this campaign we regularly confront decision makers, employing tactics of direct action. While we have not *yet* made decision makers do what we want, history shows that direct action is a powerful mechanism for creating change.

One example of a campaign that has confronted power holders was the movement to prevent the damming of the Franklin River in Tasmania, Australia from 1976–83. Drawing on the traditions of non violent direct action from Gandhi's salt marches during the liberation struggle in India, suffrage campaigns, the civil rights movement, and the anti-nuclear movement, activists blockaded construction of the dam for four months. This movement forced the Australian Federal Government to step in and overturn State Government Policy and, through legislation and a landmark court case, stop the construction of the Franklin Dam (The Wilderness Society 2006).

In ASEN's view, direct action is an essential element in the creation of circumstances whereby it is 'politically, socially, economically and physically impossible for governments to continue expanding the coal industry' (Creenaune 2008). Given Australia's coal usage and that coal is the most climate-polluting fuel, creating these circumstances is essential to stopping climate change. A recent Greenpeace report *Energy revolution: a sustainable Australia energy outlook* shows that entirely replacing coal with energy efficiency and renewable energy by 2030 will give a net gain of 10,000 jobs in the Australia economy (Greenpeace Australia Pacific 2008). Another report, by the Australian Conservation Foundation and the Australian Council of Trade Unions (2008), claims there is potential to grow an additional 500,000 green jobs in Australia by 2030. As such 'we can have jobs and an economy without coal, we can have unions, communities and families without coal, but we can't have coal without climate change' (Phillips 2007).

In addition to small direct actions, ASEN is currently involved in organizing the first Australian Camp for Climate Action. The camp is based on a model used successfully in the UK. It involves bringing together people from all walks of life to learn, live sustainably and democratically, and take collective action against the root causes of climate change (Monbiot 2008). Specifically, in Australia the Camp for Climate Action is six days of workshops and community direct action aimed at shutting down the world's largest coal port.

Many people in social movements and broader society shy away from direct action or dismiss it as radical, particularly when it involves risking arrest. I believe that direct action is one of the most responsible acts one can undertake, because 'when the laws are unjust or are destroying our future, people of conscience have a responsibility to act' (Hepburn 2007). Even prominent but arguably conservative climate advocates and scientists recognize the important role of direct action. In 2007, ex-United States Vice President Al Gore (in Kristof 2007) said 'I can't understand why there aren't rings of young people blockading bulldozers and preventing them from constructing coal fired power plants'. Dr James Hansen (2007) top scientist at The United States National Space Agency (NASA) is quoted as saying 'it seems to me that young people, especially, should be doing whatever is necessary to block construction of dirty coal-fired power plants'. For the reasons articulated in this chapter young people are very well placed to do just this: what Hansen and Gore do not seem to realize is that we have already started.

Conclusion

The Australian Student Environment Network is just one example of an organization that is creating substantive social change to avert dangerous climate change. We do not work alone. We are part of a diverse and significant social movement drawing inspiration and support from international youth networks, Australian environment groups, grassroots activists, and other social movements. In turn, we support students to become lifelong social change makers. We are successful because of our diversity of tactics and because we are not only concerned about the end goal, hence we also prioritize process. While social movements and ASEN are not perfect, they are an integral part of making a more environmentally just and sustainable society. However, it is important to note that there are key problems that ASEN faces regularly which include:

- Lack of funding or competition with other environment organizations for funding.
- High turnover of active members, which is a problem faced by youth activists' organizations worldwide.
- The political and practical tension between being a grassroots network and an organization.
- Time—there is generally not enough of it.

Nevertheless, although these problems are not insignificant they pale compared to the problem that the Earth faces in climate change. For us to have any hope of averting runaway climate change we must build a strong and diverse social movement, which includes ordinary people who are prepared to risk arrest and take direct action.

■ SUMMARY

- Social movements and direct action are important factors for creating change on climate change, particularly in Australia.

- Youth movements are key stakeholders and change makers, as illustrated by the Australian Student Environment Network (ASEN).

- ASEN uses several different change-making approaches—education, confronting power holders, influencing decision makers, and building alternative structures.

- ASEN and the broader youth climate movement face several challenges related to funding, turnover of active members, being both a grassroots network and a formally constituted organization, and time demands.

■ REFERENCES

Australian Conservation Foundation and Australian Council of Trade Unions (2008) *The Green Gold Rush*. Available from www.acfonline.org.au/uploads/res/Green_Gold_Rush_final.pdf, accessed December 2008.

ASEN (2006) *Charter and Constitution of the Australian Student Environment Network*. Available from http://www.asen.org.au/index.php?p=what_is_asen, accessed February 2006.

Australian Coal Association (2007) *Australian Black Coal Exports*. Available from http://www.australiancoal.com.au/exports.htm, accessed June 2008.

Australian Greenhouse Office (2007) *National Greenhouse Inventory 2005*. Available from http://www.greenhouse.gov.au/inventory/2005/index.html, accessed June 2008.

Creenaune, H. (2008) *Coal and the climate movement: the role of direct action*. Presentation at Curing Australia's Coal Addiction, Sydney University, June 2008.

Coover, V., Deacon, E., Esser, C. and Moore, C. (1977) *Resource Manual for a Living Revolution*. Philadelphia, PA: New Society Press.

Department of Climate Change (2007) *Greenhouse Gas Emissions Projections 2007 Stationary Energy Sector*. Available from http://www.climatechange.gov.au/projections/pubs/energy2007.pdf, accessed September 2008.

Diesendorf, M. (2007) *Greenhouse Solutions with Sustainable Energy*. Sydney: UNSW Press.

Freire, P. (1970) *Pedagogy of the Oppressed*. New York: Continuum Publishing.

Greenpeace Australia Pacific (2008) *Energy Revolution: A Sustainable Australia Energy Outlook*. Available from http://www.greenpeace.org.au/energyrevolution/landing.php?source=Index, accessed September 2008.

Hamilton, C. (2007) *Scorcher: The Dirty Politics of Climate Change*. Melbourne: Black Inc. Books.

Hansen, J. (2007) Old King Coal, Climate Progress blog, accessed June 2008, http://climateprogress.org/2007/07/06/james-hansen-on-stopping-new-coal-plants/

Hepburn, J. (2007) The Camp for Climate Action. Email on the climatecamp_communications@lists.riseup.net elist, 9 October 2007.

Highlander Center (Date Unknown) *Highlander Research and Education Center—About Us*. Available from http://www.highlandercenter.org/about.asp, accessed August 2008.

Kristof, N. D. (2007) The big melt. *The New York Times* 16 August. Available from http://select. nytimes.com/2007/08/16/opinion/16kristof.html?scp=2andsq=the+big+meltandst=nyt.

Monash Climate Change Group (2005) *Proposal for how Monash can Reduce its Contribution to Climate Change*. Available from http:// www.asen.org.au/index.php?p=campaign_clean_ energy, accessed June 2006.

Monash University (2005) Monash to reduce energy consumption by 20% by 2010. *Monash Newsletter* 15 October. Available from http://www.monash.edu/news/newsline/story/632.

Monbiot, G. (2008) *The discrepancy between climate policy and practice*. Curing Australia's Coal Addiction, Sydney University, June 2008.

Moyer, B., McAllister, J., Finley, M-L. and Soifer, S. (2001) *Doing Democracy: The MAP Model for Organising Social Movements*. Gabriola Island: New Society Publishers.

Pearse, G. (2007) *High and Dry: John Howard, Climate Change and the Selling of Australia's Future*. Melbourne, Australia: Penguin.

Phillips, S. (2007) Speech to the Newcastle Coal Port flotilla blockade, Newcastle, November 2007.

Ramachandran, A. (2007) 11 in coal protest. *Sydney Morning Herald*. Available from www.smh. com.au/news/environment/eleven-held-in-coal-protest/2007/09/04/1188783200074.html, accessed September 2007.

Scott, J. and Marshall, G. (2009) *A Dictionary of Sociology*. Oxford: Oxford University Press. Available from Oxford Reference Online http://www.oxfordreference.com/views/ENTRY. html?subview=Main&entry=188.e2148, accessed May 2009.

Sourcewatch (2008) *Non-violent Direct Action Against Coal*. http://www.sourcewatch.org/index. php?title=Nonviolent_direct_actions_against_coal.

The Wilderness Society (2006) *History of the Franklin River Campaign, 1976–1983*. Available from http://www.wilderness.org.au/articles/franklin, accessed September 2008.

Turton, H. (2004) *Greenhouse Gas Emissions in Industrialised Countries: Where does Australia stand?* Australia Institute discussion paper accessed June 2008 http://www.tai.org.au/documents/ downloads/DP66.pdf.

United Nations Development Program (2007) *Human Development Report—China*. Available from hdrstats.undp.org/countries/data_sheets/cty_ds_CHN.html, accessed September 2008.

■ USEFUL WEBSITES

For more on green jobs and climate solutions:

ACF and ACTU (2008) *The Green Gold Rush*. Available from **http://www.acfonline.org.au/uploads/ res/Green_Gold_Rush_final.pdf,** accessed December 2008.

Diesendorf, M. (2007) *Greenhouse Solutions with Sustainable Energy*. Sydney: UNSW Press.

Greenpeace Australia Pacific (2008) *Energy Revolution: A Sustainable Australia Energy Outlook*. Available from **http://www.greenpeace.org.au/energyrevolution/landing.php?source=Index,** accessed September 2008.

Jones, V. (2008) *The Green Collar Economy: How One Solution Can Fix Our Two Biggest Problem*. New York, NY: HarperOne.

For more on popular education and theories of change:

Coover, V., Deacon, E., Esser, C. and Moore, C. (2007) *Resource Manual for a Living Revolution.* Philadelphia, PA: New Society Press.

Freire, P. (1970) *Pedagogy of the Oppressed.* New York: Continuum Publishing.

Highlander Center **http://www.highlandercenter.org**

The Change Agency **http://www.thechangeagency.org/**

For more on the Australian coal industry:

Hamilton, C. (2007) *Scorcher: The Dirty Politics of Climate Change.* Melbourne: Black Inc. Books.

Pearse, G. (2007) *High and Dry.* Penguin Australia, Australia.

For more on the climate movement and direct action:

Australian Student Environment Network **http://www.asen.org.au**

Climate Camp Australia **http://www.climatecamp.org.au**

Climate Camp UK **http://www.climatecamp.org.uk**

It's Getting Hot in Here: Global Dispatches from the Youth Climate Movement **http://itsgettinghotinhere.org/**

Sourcewatch (2008) *Non-violent Direct Action Against Coal* **http://www.sourcewatch.org/index. php?title=Nonviolent_direct_actions_against_coal**

8

Social dynamics of the US environmental challenge

Ben Crow

Introduction

Environmental views in the US, as in other industrial countries, are divided between optimists and pessimists. The optimists believe that human creativity under capitalism can find technological solutions to the challenges of environmental sustainability. Pessimists are concerned that solutions will come too late, and portray a dark future of environmental decline. Whether the optimists or the pessimists are right depends largely on the outcomes of contending processes of social change.

In this chapter I explore some of the social and intellectual dynamics of US environmentalism. This exploration suggests first that our ability to reflect upon who we are and what we do, means that how we understand our relation to nature helps form that relationship. Secondly, it implies that the social dynamics of capitalism, its ability to sponsor and embody creativity as well as its social divisions of class and race, are central to our understanding of what can be done.

Reflexivity and the debate about wilderness

Wilderness preservation is a key goal of the US network of national parks (Figure 8.1), an array of stunning landscapes which are globally renowned. Wilderness in this sense is a natural environment that has not been modified by human activity. Within many national and local parks, significant areas specified as 'wilderness' are kept untouched by roads and other intrusive human activity.

In a controversial and influential 1996 essay, environmental historian William Cronon questioned the central place of wilderness in US thinking about the environment. The essay described the historical currents which shaped reverence for wilderness and

Figure 8.1 US National Parks.

suggested how this focus could constrain action and understanding. The essay provided an excellent example of our reflexive nature — our ability to reflect upon ideas and actions and to respond to that reflection.

Cronon called his essay 'The trouble with wilderness: or getting back to the wrong nature'. He wrote:

> [the] modern environmental movement is itself a grandchild of romanticism and post-frontier ideology, which is why it is no accident that so much environmentalist discourse takes its bearings from the wilderness these intellectual movements helped to create . . . the wilderness serves as the unexamined foundation on which so many of the quasi-religious values of modern environmentalism rest.
>
> Cronon (1996: 72, 80)

The essay raised prolonged debate partly because it suggested that wilderness preservation should not be the central focus of environmental organizing.

Cronon's essay noted the long-standing and widespread human influence on the US landscape and situated reverence for wilderness at the confluence of two historically constructed ways of thinking, or **discourses**.

- Romanticism, which originated in Western Europe in the late 18th century and spread to America, placed a high value on untamed nature—considered to be sublime, awe-inspiring, and of the highest spiritual value—as part of a reaction against the industrial revolutions of that epoch. It was epitomized in the writing of Wordsworth, Thoreau, and Muir. Cronon thus traced how the romantic movement had transformed the idea of wilderness from a kind of hell, desolate, barren, and deserted places where Christ, for example, famously faced the devil, into heaven on earth, a sacred place where untouched nature can be revered, 'the last remaining place where civilization . . . has not fully infected the earth' (Cronon 1996: 69).

- US frontier ideology, which was a uniquely American set of ideas, saw the engagement of settlers with untouched lands and survival in primitive conditions as a crucible of American identity.

Wilderness thus became 'the quintessential location for experiencing what it meant to be an American' (Cronon 1996: 76).

> *William Wordsworth* (1770–1850) was an English poet who helped launch the romantic age in English literature. His autobiographical poem *The Prelude* describes the almost religious awe he experienced climbing in the European Alps.
> *Henry David Thoreau* (1817–1962) was an early American naturalist writing on the environment, simple living, and resistance to government.
> *John Muir* (1838–1914) was an American naturalist best known for a campaign to conserve the Yosemite valley in California.

So, what is 'the trouble with wilderness'? It creates too simple a dichotomy between nature and humans, claiming undivided attention for the playgrounds particularly of the privileged, while ignoring the polluted locations of the poor.

> *The wilderness dualism tends to cast any use as ab-use, and thereby denies us a middle ground in which responsible use and non-use might attain some kind of balanced and sustainable relationship . . . only by imagining this middle ground will we learn ways of imagining a better world for all of us: humans and nonhumans, rich people and poor, women and men, First Worlders and Third Worlders, white folks and people of color.*

> *Cronon (1996: 85–6)*

Cronon's essay also argued that human influence on the environment was more pervasive and extensive than commonly understood. Areas designated wilderness often incorporated areas which had been shaped by humans. The creation of wilderness also frequently required the exclusion of previously established human activity. As history professor Karl Jacoby said, 'There weren't empty wilderness areas in the United States. They had to be created by the removal of Indians' (Doyle 2008: 23).

A response to Cronon came from another environmental historian, Donald Worster (1997). He argued that Cronon, and like-minded colleagues, had made three errors:

1. asserting that America was never a wilderness,

2. arguing that wilderness was only a cultural construct of rich white romantics,

3. saying that wilderness preservation had distracted from other more pressing environmental concerns.

Worster argued that 2 million native Americans could not have domesticated the whole of north America; that love of nature was not simply a 'cultural construct' but 'may even have roots in the very structure of human feelings and consciousness' (1997: 11), and that the wilderness movement has not distracted attention from other, more mundane environmental questions.

This debate continues. From a sociological perspective, it emphasizes two points: first, the historical contribution of ideas from both high culture (the arts depicting nature as sublime) and popular imagery (the image of rugged individualism in settling the American frontier) to how humans in general and Americans in particular think about nature; and second, how the way humans think about nature has implications for how we relate to nature. The transition in the meaning of wilderness, from hell to the sublime, led to the setting aside of wilderness areas and national parks. This is reflexivity in practice: humans think about, and in the process change, themselves and their surroundings.

The wilderness debate in the US also has practical implications in the developing world. National parks, modelled on US national parks, have arisen in many countries. Ramachandra Guha, an environmental historian based in India, has argued that the establishment of parks 'results in a direct transfer of resources from the poor to the rich' (Guha 1989: 75).

This debate about the place of wilderness in US environmental thinking questions the way that reverence for wilderness has generated an environmentalism primarily focused on protected parks and wild areas. Cronon and others suggest that a broader concern for everyday interactions with nature, in cities, slums, and suburbs, could generate a more inclusive movement toward sustainability, an environmentalism that is also for the poor and the global south.

Race, class and environmental justice

The obverse face of US national parks is the proliferation of areas polluted by military and industrial toxic waste in hazardous dumps spread across the country. A 'superfund' was established by legislation in 1980 to finance the clean-up of these sites. In retrospect we can see that the legislation underestimated the difficulty and cost of environmental clean-up. Despite considerable expansion of the fund, the task of making more than 1,000 sites safe has hardly begun. And these sites are but a small fraction of the hazardous waste dumps—estimated to be up to 425,000 (US General Accounting Office 1987, cited in Hird 1993: 323)—produced by industrial and military activity.

Many of these superfund sites and other hazardous waste dumps are in, or adjacent to, poor and minority communities (Brown 1995: 29). Activism and research have begun to illuminate how the normal workings of the US economy tend to situate toxic waste sites and poor and minority communities adjacent to one another. This is a social dynamic which exacerbates divisions of race and class, and is the focus of an important social movement.

The Environmental Justice movement deployed methods of organizing from the civil rights movement to counter the injustice of the location of toxic waste and polluting sites. Thus, the internet-based encyclopaedia, Wikipedia, describes Environmental Justice as:

a movement that began in the U.S. in the 1980s and seeks an end to environmental racism. Often, low-income and minority communities are located close to highways, garbage dumps, and factories, where they are exposed to greater pollution and environmental health risk than the rest of the population. The . . . movement seeks to link 'social' and 'ecological' environmental concerns, while . . . keeping environmentalists conscious of . . . racism, sexism, homophobia, classism, and other malaises of dominant culture.

Wikipedia (2009)

The charge of environmental racism implies that toxic waste sites and ongoing industrial pollution sources were located close to minority communities as a result of prejudice and discrimination. The charge is sometimes justified but careful review of the question (Szasz and Meuser 1997) documents a more complex picture in which the normal workings of the capitalist markets for land and labour may be more significant than conscious choices directly involving racism. Toxic waste sites lower land values, and polluting industrial facilities seek cheap land. Low-paid minorities and working people are forced in addition to seek residence in areas with lower prices.

Rather than attributing the location of toxic wastes close to minority and poor communities to racism, Szasz and Meuser suggest that environmental inequality is a feature of modern societies which:

transform nature to a quantitatively and qualitatively unprecedented degree . . . those transformations of nature have adverse human/social impacts [which] fall unevenly along existing divisions of wealth/poverty, power/powerlessness . . . [and this in turn tends to] reproduce and exacerbate existing social inequalities.

Szasz and Meuser (1997: 116)

Thus, sustainability in the US may require greater economic equality and more effective representation of poor and minority communities. Economic trends have, for at least the last decade, been pointed in the opposite direction. There has been rising economic inequality (Duménil and Lévy 2004). In the next section, I examine technological optimism which suggests that changes in the regulation of capitalism may be sufficient to reduce US output of greenhouse gases.

Technological optimism and the dynamics of the US economy: global warming and oil

Will the US economy respond to the challenge of global warming? Can human creativity be channelled into technological innovation through existing corporate and governmental institutions? There is a debate about these questions in which technological optimists respond affirmatively to both questions (Cox 2004).

In this section, I describe a set of ideas suggesting that US oil use can be halved in twenty years, and the other half replaced by alternatives to oil, and, in the process, corporate profits can increase. If oil use can be reduced in this way then US emissions of global warming gases can be substantially reduced. These ideas provide an example of technological optimism. This strand of thinking suggests that human creativity can be embodied in new technologies whose adoption is encouraged by cost-saving in businesses and innovative regulation. Other examples of technological optimism are described in Benton (1994), Hawken, Lovins, and Lovins (1999), Wallace (1996), McDonough and Braungart (2002) and Lomborg (2001, 2004).

While industrial consumption remains the largest user of energy, transportation uses nearly as much, according to the Quick Start partnership of US industry, government and non-profit organizations (http://www.energyquickstart.org/). The ideas I am describing focus particularly, but not entirely, on oil use in transport. Parallel possibilities are thought to exist in the use of energy in industry.

A 2004 report, partly funded by the Pentagon, expressed an optimistic view about reducing US dependence on foreign oil and on oil altogether. *Winning the Oil End-Game* (Lovins *et al.* 2004) outlined a set of business friendly innovations for a two-decade transition from oil to alternative energy sources. The book and popular articles (e.g. Lovins 2004) described routes enabling half of US oil use to be conserved, through greater efficiency, and the other half to be replaced by alternatives.

Lovins and colleagues proposed a series of measures, including:

1. *Doubling the efficiency of oil use*: energy used in transport, the report suggested, can be sharply curtailed through the use of very light vehicles powered by hybrid (combined petrol and electric) energy.

2. *Encouraging adoption of more efficient vehicles* including: 'feebates' (fees on inefficient vehicles and rebates for the efficient); a scrap and replace programme (where government buys gas-guzzling old cars and leases efficient cars to replace them); loan guarantees for the airline and auto industries introducing new technologies.

3. *Replacing oil with plant energy*: use of plants as a substitute for fossil fuels.

The authors suggested that the transition from oil to renewable energy, and lower emissions of greenhouse gases, could be profitable at the level of oil prices then prevailing, US$26 per barrel. In an article in the leading US business magazine *Fortune*, Lovins described the set of actions which could realize their proposals:

> *Astute business leaders can turn innovation from a threat to a friend. Military leaders can support advanced-materials R&D and procure super efficient platforms. Political leaders can craft policies that stimulate demand for efficient vehicles, reduce investment risks, and purge perverse incentives. Citizens must play a role too—a big role—because their choices guide the markets, enforce accountability, and spur social innovation. The surprise popularity of Toyota's Prius, Honda's Civic hybrid, and Ford's Escape hybrid . . . suggest that consumers welcome efficient designs if they're appealing.*
>
> *Lovins (2004: 3)*

At mid-2008 oil price levels of over US$100 per barrel, the potential for corporations to fund the transition with savings in energy costs are enhanced. (In late 2008, the US government provided US$25 billion in loans to hasten this transition.) We should note, in passing, the light social analysis in Lovins' description. Technology and business, it is assumed, will lead the change, with citizens cast as exercising influence only through their role as consumers of the products of industry.

There is some evidence that past price rises have resulted in transient reductions in oil consumption. During the oil price hikes of 1974 and 1979, transport and industrial energy consumption showed some sign of levelling off or declining. In 2008, again, the rising price of oil began to have noticeable impact on private transport: gas-guzzlers became more difficult to sell and the distances people travelled by car began to fall.

There are incentives within capitalism that have sometimes encouraged a transition to greater energy efficiency. Wallace (1996: Chapter 4) describes the greater energy and material efficiency of successive waves of industrial organization, from the artisanal industrial production of the English industrial revolution in the late 1700s, to Fordist mass production in the US in the early twentieth century, to the flexible or lean production emerging in Japan after the Second World War.

As energy costs rise, the incentives to save money by reducing energy use in all aspects of production and consumption increase. Indeed, the efficiency of energy used in the US, the UK, Japan, and China has been increasing steadily. The US is, nonetheless, a laggard by this measure, behind even China since about 1998, in the efficiency of energy use in production. Meanwhile, Goldemberg and colleagues (reproduced in Wallace 1996) suggest that this pattern of increasing energy efficiency holds across nations, where successive industrializing nations tend to leapfrog ahead with greater energy efficiency than those nations that industrialized earlier.

Rising energy *efficiency* should not, however, be misread as a description of falling energy use. Increases in energy efficiency are overwhelmed in all of these countries by the ongoing rise in economic output. Even though less energy is used per unit of production, substantially more goods and services are being consumed. So, the overall use of energy is increasing. This suggests we need to know more about the dynamics of energy consumption (Shove 1997, 2003) and growth of economic output. I do not deal with these questions here.

Critics of the technological optimists counter that the relocation of energy-intensive manufacturing from industrialized countries, like the US and Japan, to low-wage, newly industrializing countries, notably China and India, may contribute to a fall in energy efficiency in the former. Thus Vaclav Smil (2007) has shown that major industrial sectors in Japan started to use more energy after 1990, even as the energy efficiency of the whole economy continued to improve. This happened as production in those industries was moved to China. Smil suggests that the relocation of manufacturing industry to China reduced competitive pressure driving energy efficiency in plants that remained in Japan. What happened to energy efficiencies in the industries relocated to China needs to be investigated.

Technological optimism is controversial and these ideas need much more analysis (Press 2007). As the label suggests, technological optimists tend to be stronger on harnessing innovation through technical change than identifying and assessing the social relationships, institutions, and practices assumed to disseminate change. Wallace (1996) suggests that the momentum of existing institutions will limit the pace of change in industrialized countries and that sustainable social systems are more likely to emerge in the newly industrializing world.

Environmentalism and the social dynamics of environmental change

[E]nvironmentalism will never be able to muster the strength it needs to deal with the global warming problem as long as it is seen as a 'special interest.' And it will continue to be seen as a special interest as long as it narrowly identifies the problem as 'environmental' and the solutions as technical.

Shellenberger and Nordhaus (2004: 26)

In 2004, a corporate consultant, Michael Shellenberger, and a pollster, Ted Nordhaus, published an essay called *The death of environmentalism* which was widely noticed and elicited considerable response from the US environmental movement. Their essay, which they followed with a book in 2007, argued that the US environmental movement had lost its way, 'modern environmentalism is no longer capable of dealing with the world's most serious ecological crisis [global warming]' (Shellenberger and Nordhaus 2009: 6). Their argument has parallels with Cronon's in the debate about wilderness. Shellenberger and Nordhaus argue that the insistence by leaders of the environmental movement that nature is a thing, separate from humans (the reification of nature), makes it difficult to build political coalitions with powerful potential allies like the labour movement and sectors of industry.

They suggest, for example, that the repeated failure to get legislation passed limiting the fuel usage of cars and trucks arose from opposition of labour unions in the automobile industry. The environmental movement's inability to generate a vision of social change which went beyond the environment to include jobs meant that the auto industry unions would not support the legislation.

This brings us to the notion of sustainability in the US which, as elsewhere, involves ideas about the environment, struggles over inequality and the economy, and assessments of the ability of more or less regulated capitalism to achieve appropriate innovations.

Is the environment constituted by wilderness, nature unsullied by humans, or is that a dream of past romance? Is it alternatively constituted in the cities and countrysides by human interactions with their everyday world? These two visions suggest different perspectives on and priorities for sustainability. The first is easy to imagine, the second more complex. If humans are part of nature, what do we sustain?

Land prices encourage hazardous waste sites to be situated near the poor and excluded. While their voices are under-represented, economic growth may be underpinned by the easy disposal of waste and a proliferation of toxic hazards. If, however, they gain influence through electoral politics and movements like environmental justice, then regulatory and other constraints may limit heedless industry.

Concern about toxic hazards and global warming contributes to the regulation of capitalism. Regulation shapes innovation. The growth of widespread fears about toxicity and climate change has contributed to the spread of green consumerism, building, industry, and cities. But views about the ability of even a regulated capitalism to respond adequately are sharply divided.

▓ SUMMARY

- American ideas on the environment are shaped by the notion of wilderness, with its origins in romanticism and the idea of the frontier.

- An environmentalism that situates human activity as an integral part of nature will identify different priorities to one focused on wilderness.

- Social inequalities refract economic processes to generate intolerable injustices in living conditions, which more effective representation could overcome.

- Technological optimists foresee profitability, backed by regulatory incentives, as a significant influence on a transition to sustainability.

- In the US, as elsewhere, the forces that shape change toward sustainability include ideas, social movements, corporate innovation, the powers of government, and the economic and social power of capitalism.

▓ ACKNOWLEDGEMENTS

My colleague Andrew Szasz introduced me to the Cronon–Worster debate and over the years has been a significant influence on my thinking about the environment. Another colleague, Ali Shakouri, has contributed to my understanding of energy use through our in-class debates on the topic. He also drew my attention to the findings of Vaclav Smil on energy efficiencies for major industrial sectors in Japan. Students of my class Sociology 179 Nature, Poverty and Progress have also provided constructive criticism. Neither my students nor my colleagues, however, are implicated in my formulation of the issue.

▓ REFERENCES

Benton, T. (1994) Biology and social theory. In M. Redclift and T. Benton (eds), *Social Theory and Global Environment*, pp. 28–50. New York: Routledge.

Brown, P. (1995) Race, class and environmental health: a review and systemization of the literature. *Environmental Research* 69: 15–30.

Cox, S. (2004) From here to the economy: can capitalism be harnessed to solve environmental problems, or is capitalism itself the problem? *Grist online journal*. Available from http://www.grist.org/article/cox-economy

Cronon, W. (1996) The trouble with wilderness: or getting back to the wrong nature. In W. Cronon (ed.), *Uncommon Ground: Rethinking the Human Place in Nature*, pp. 69–90. New York, NY: Norton.

Doyle, L. (2008) Sitting Bull's tribe to regain control of southern Badlands. *The Independent* (London): 10 June, 23.

Duménil, G. and Lévy, D. (2004) Neoliberal income trends: wealth, class and ownership in the USA. *New Left Review* 30: 105–133.

Guha, R. (1989) Radical American environmentalism and wilderness preservation: a Third World critique. *Environmental Ethics* 11(Spring): 71–76, 78–83.

Hawken, P., Lovins, A. and Lovins, H. (1999) *Natural Capitalism: Creating the Next Industrial Revolution*. New York: Little, Brown and Co. and London: Earthscan.

Hird, J. A. (1993) Environmental policy and equity: the case of superfund. *Journal of Policy Analysis and Management* 12(2): 323–343.

Lomborg, B. (2001) *The Skeptical Environmentalist*. Cambridge: Cambridge University Press.

Lomborg, B. (2004) *Global Crises, Global Solutions*. Cambridge: Cambridge University Press.

Lovins, A. (2004) How America can free itself of oil—profitably. *Fortune* 4 October: 1–3. Available from http://www.rmi.org/sitepages/pid171.php#OilDependence.

Lovins, A., Datta, E., Bustnes, O. E. and Koomey, J. (2004) *Winning the Oil End-Game*. Boulder, CO: Rocky Mountain Institute.

McDonough, M. and Braungart, W. (2002) *Cradle to Cradle: Remaking the Way We Make Things*. New York: North Point Press.

Press, D. (2007) Industry, environmental policy and environmental outcomes. *Annual Review of Environment and Resources* 32: 317–44.

Shellenberger, M. and Nordhaus, T. (2005) The death of environmentalism: Global warming politics in a post-environmental world. Available from http://www.grist.org/news/maindish/2005/01/13/doe-reprint/.

Shove, E. (1997) Revealing the invisible: sociology, energy and the environment. In M. Redclift and G. Woolgate (eds), *The International Handbook of Environmental Sociology*, pp. 261–272. Cheltenham, UK and Northampton, MA: Elgar.

Shove, E. (2003) *Comfort, Cleanliness and Convenience*. Oxford, New York: Berg.

Smil, V. (2007) Light behind the fall: Japan's electricity consumption, the environment, and economic growth. *Japan Focus* 99. Available from http.//www.japanfocus.org/-Vaclay-Smil/2394.

Szasz, A. and Meuser, M. (1997) Environmental inequalities: literature review and proposals for new directions in research and theory. *Current Sociology* 45(3): 99–120.

US General Accounting Office (1987) Superfund: extent of nation's potential hazardous waste problem still unknown. Washington, DC: Government Accountability Office.

Wallace, D. (1996) *Sustainable Industrialization*. London: Royal Institute of International Affairs and Brookings Institution.

Wikipedia (2009) http://en.wikipedia.org/wiki/Environmentla_Justice, accessed 11 January 2009.

Worster, D. (1997) The wilderness of history. *Wild Earth* Fall: 9–13.

▉ FURTHER READING

Cronon, W. (1983) *Changes in the Land: Indians, Colonists, and the Ecology of New England*. New York: Hill and Wang, 1983; 20th anniversary edition, 2003.

McDonough, W. and Braungart, W. (2002) *Cradle to Cradle: Remaking the Way We Make Things*. New York: North Point Press.

Lovins, A., Datta, E., Bustnes, O. E. and Koomey, J. (2004) *Winning the Oil End-Game*. Boulder, CO: Rocky Mountain Institute.

Szasz, A. (2007) *Shopping Our Way to Safety: How We Changed from Protecting the Environment to Protecting Ourselves*. Minneapolis, MN: Minnesota University Press.

9

Looking beyond the visible: contesting environmental agendas for Mumbai's slums

Dina Abbott

Introduction

A striking growth in the urban populations of southern mega–cities (with populations exceeding 8 million) and hyper–cities (over 20 million) is accompanied by the depressing phenomenon of ever-expanding slums, especially since the 1970s (Davies 2006). In fact, it is becoming increasingly difficult to demark urban and peri-urban areas because public land, whether within the inner or outer circles of cities, is encroached by a variety of slum developments. Millions of people live in these slums as they are pushed out of rural areas towards cities which promise livelihood and income opportunities, even if these are as undignified as rummaging through the city's waste. As rural ties break down, and there is little hope of returning, generations continue to reside in the most squalid and appalling inhuman conditions of urban slums.

Peri-urban areas are those that develop haphazardly, fringing on and/or between official urban and rural municipal and administrative boundaries. As the city expands, peri-urban areas grow with it. This process is often so rapid that areas previously defined as 'peri-urban' soon get integrated into the 'urban', making it very difficult for official mapping, data collection, and enforcement of regulations. An important factor in the growth of specific peri-urban areas is how close they are to the nearest urban centres and what social and economic opportunities they offer.

Slums are the most immediate, visible symbols of poverty and environmental degradation intertwined in cities. They are a constant reminder of national shame and the state's incapacity or political will to tackle poverty. In cities where the poor and rich share spaces, the rich will attempt to mentally and morally distance themselves from the slums, often regarding these as eyesores, health hazards, and dens of corruption and immoral behaviour. Yet slums are home to millions, from single householders to intergencrational extended families. Within each slum locality, there is intense social networking to safeguard common interest, provide informal services for neighbours and enhance the ability to carry out livelihood opportunities.

There is therefore a clear contrast in the way slums are regarded by 'outsiders', and those who actually live there. Equally there is a difference in which both outsiders and slum dwellers understand environmental needs. For instance, whilst slum dwellers may seek a 'brown' agenda that enables better access to basic services, health, and long-term security, the 'green' agenda often led by outsiders revolves around pleasant urban spaces that enhance views, provide cleaner public spaces, and in turn increase property values of neighbouring localities. This results in a continuous battle between those who live in the slums and those who do not in attempting to define environmental agendas, sometimes escalating into violent dispute. Success depends on each group's power to negotiate with state authorities and state representatives, its social ranking, and its relationship with richer neighbours.

The green agenda is associated with broad, conventional environmental concerns of climate change, biodiversity, resource depletion, deforestation, and conservation. The brown agenda applies particularly to livelihood concerns in cities, such as overcrowding, inadequate housing, access to clean water and sanitation, and deteriorating air quality (Beall 2002). Because green agendas are often associated with long-term, wider concerns, contrasting with brown agendas that focus on immediate and localized issues, these are often seen as contrasting, even conflicting. However, both centre on environmental interests and are interdependent and interlinked. For instance, an individual's fight for the right to proper sanitation services can be regarded as a localized brown issue, whereas this is in fact also a part of the wider green agenda which is concerned about pollution, health hazards, and the amelioration of global poverty.

A key question for this chapter is, therefore, what is the contested nature of environmental agendas in urban areas and who or what defines it? To answer this question, I will draw on examples from Mumbai (see Figure 9.1) to argue that within shared spaces, whilst there may be commonality of environmental interests, environmental agendas are often shaped by those who are more powerful and vocal.

Historically, Mumbai, the capital of the State of Maharashtra, has been one of India's most important cities and ports. Today too it continues to dominate commerce, finance, trade, and media (including the large Bollywood film industry). It is where the money is to be made and is therefore a magnet for both the rich and the poor alike, creating huge problems associated with daily uncontrolled in-migration and overcrowding, defeating the Indian government's continuous effort to make Mumbai a world-class city.

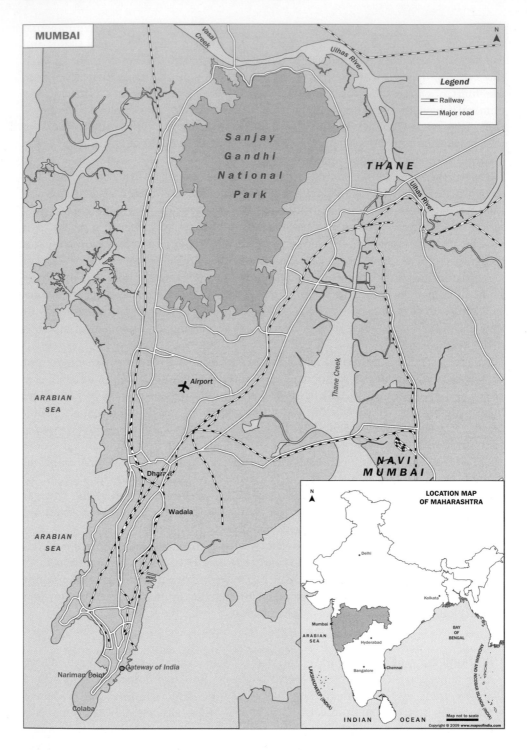

Figure 9.1 Location map of Mumbai, capital of the State of Maharashtra.

What are slums?

It is important to deconstruct the category 'slums'. This is because slums are not homogenous, and in reality there is a vast differentiation between slum types, their geographies, histories, and peoples who live here. Slums are usually known by their regional and local variations, for example, as bustees (in Kalkota, Bangladesh), favelas (in São Paulo), and shanty towns (Johannesburg, Nairobi). They usually lack sanitation, are bursting at the seams, and their dwellings can be found in every conceivable public place—for example cemeteries in Cairo and drainpipes in Mumbai—where individuals attempt to establish personal spaces and boundaries.

In Mumbai, where the population of the city (excluding Greater and New Mumbai) exceeds 13 million, at least 55 per cent live in slums (Government of India 2001). In fact, all over India such large numbers now live in slums that the Indian government, for the first time, incorporated a special category to enumerate slum dwellers in the 2001 census.

Whatever the term, it is important to recognize that all slum localities have their own histories, social, and political contexts, and differences in the legality of their existence. For example, in Indian states, hugely complex central government and state government legislation going back to colonial times decides what makes or does not make a slum. This legislation has undergone several amendments and changes in response to criticism over the years. I do not have the space to discuss the intricacies of these here, so I will refer to the Maharashtra Slum Areas (Improvement, Clearance and Redevelopment) Act of 1971 which covers Mumbai directly, and claims to be the most comprehensive, inclusive piece of legislation. From time to time, based on loose definitions (such as congested, unhygienic, public hazard) of what makes a slum, certain localities are officially declared slums under the Act.

Official recognition of this type is very important for the survival of the slum, as it means that this slum is on its way to becoming an established residential area (Abbott *et al.* 1998a). Theoretically, once recognized, the state government is obliged to supply the slum with basic services of water supply, drainpipes, sanitation, and sometimes electricity. However, it is not in the interest of planners, politicians, private developers, and ultimately state governments to provide basic services for more than a handful of slum pockets. Therefore, those that have not officially been 'declared' continue to be denied access to water and sanitation, and are constantly threatened with demolition. Also, it is worth noting that it is not unusual to find newer settlements springing up around the edges of recognized slums so that they can gain illegal access to basic services found there. The overall effect of this is that boundaries between what is official and what is not quickly become blurred and are rarely modified. Therefore, it is quite common to find that even those pockets which have managed to negotiate legal backing, and are entitled to some basic protection from the state, are very soon stretched to the limit.

In Mumbai, the oldest of slum housing can be traced back to colonial times in tenement housing, known as 'chawls' (Figure 9.2). Chawls were originally constructed to house the many seasonal rural migrants who worked for the British East India Company in docking,

Figure 9.2 A 'chawl' building, Mumbai.

shipping, and other burgeoning activities of Mumbai's textile mills around the 1800s. Throughout the centuries chawls have seen modifications and change of ownership, especially during disease outbreaks such as that of the bubonic plague in the 1890s, late 1920s, and even 1990s (Burnett-Hurst 1925; Upadyay 1990; Zachariah 1968). Essentially, however, chawl structures continue to resemble the tenement blocks of Victorian Glasgow in the UK. Even today they remain dominated by people who are intergenerationally linked with migrants from neighbouring areas. In patterns established long ago, migrant urban families retain allegiance to rural areas they originated from, as is evident in the constant movement between relatives who move in and out of residential arrangements in the older established chawls.

Intense all-India rural poverty in recent times has, however, seen a significant change in patterns of migration. Instead of seasonal migration from nearby, Mumbai now has whole, destitute households arriving from far-away parts of India everyday. These families do not have historical ties with Mumbai and are forced to set up home anywhere they can in illegal, temporary, unplanned structures of plastic, corrugated iron, cardboard, and flattened tins known as 'zopadpattis' or hutments (Figure 9.3). In Mumbai, practically all unused municipal land, inner city railway lines and bridges, public as well as some large privately owned construction working sites are usurped every day by new arrivals. Because of the illegal nature of zopadpatti settlements, as well as a historical lack of support systems, the millions who live here lead precarious lives. Apart from a constant fear of eviction from city authorities, rival groups of settlers and goondas (equivalent to the Italian Mafiosi) of the localities who thrive on the vulnerability of the poor, frequently terrorize zopadpatti

Figure 9.3 Zopadpattis in Mumbai.

dwellers. Violence, drugs, and sexual assaults are common. Thus whilst chawls have legislative protection as established poor residential localities, zopadpattis are illegal and Mumbai authorities have several times acted brutally to destroy these and evict people.

Whilst there are differences between slum types, there are also differences within slums and slum hierarchies. For instance, higher-caste Hindus who have historical links with Mumbai are able to negotiate chawl housing that is closer to the centre, has better access to transport, livelihood opportunities, municipal schools, and subsidized government ration shops. At the bottom of the scale are the several thousand lower-caste Hindu converts to Buddhism and Islam who changed religion in order to escape their lowly position in Hindu society, but who have not quite managed to escape from associated stigmas. They are often excluded from chawls and live further out in unstable set-ups on the margins of established slums, sometimes on footpaths (Abbott 1993; Abbott *et al.* 1998a, b).

Thus, slums are differentiated by their legal status and by their specific histories. Slums have their own social hierarchies which also reflect power and powerlessness within groups who live there.

Sharing space: contesting agendas

Another characteristic of slums is that they share spaces with the rich in one way or another, although to differing extents. For example, in India's political capital Delhi, the rich dominate areas of the beautiful Lutyens-designed New Delhi, almost segregating the poor to the run-down parts of Old Delhi. Mumbai, however, is special in that the rich and poor live side by side, sharing public spaces and utilities. This phenomenon is evident on the first journey from the international airport to the centre of Mumbai, which weaves through Dharavi, Asia's largest slum. The sight that greets even the most frequent and hardened traveller to Mumbai is daunting and incomprehensible. For here is the financial capital of India, with miles and miles of ramshackle squalid hutments of Dharavi dotted with plush high-rise business blocks of the gold and diamond merchants. Yet this is to some extent a mutually beneficial relationship, with the poor providing the rich with many cheap informal services whilst the rich provide patronage and income opportunities (however negligible) in return. It is also one that is convenient to the rich who have cheap labour and services at their doorstep with little built-in travel time or costs.

This proximity of the rich and poor means that concerns about immediate environments are important for both groups. However, there is little doubt that the huge numbers of homeless who sprawl on city streets, encroaching public spaces in their thousands, create environmental hazards and degradation through the sheer demands made on these spaces. This creates conflict. Whilst the poor are desperate for shelter and basic services such as water and sanitation, the neighbouring rich want to demolish slums that they regard as environmental and social disasters.

In addition, there are outsider groups who also have a vested interested in the environments of various localities. For instance, Mumbai municipality will protect legal slums whilst demolishing illegal structures. Politicians are mainly concerned with the vast bank of poor votes as well as those of the rich and powerful (including the slum Mafiosi). They will therefore woo the poor with populist housing policies whilst at the same time promising to beautify Mumbai for the rich. The rich further have vested interest in property markets. They are also concerned to attract business, particularly foreign investment, all of which require clean environments and ready infrastructure. Then there are NGOs and other campaigning bodies (often backed by Bollywood superstars) who attempt to negotiate with officialdom on behalf of the poor. Yet they too appear to have underlying agendas of securing well paid (often foreign-funded) jobs or publicity for themselves. The poor therefore have to struggle hard to make their voices heard within this web of agendas.

The ability to manipulate often conflicting agendas depends on several factors including economic and social factors such as levels of income, caste, religion, gender, and political allegiance. There are numerous examples where the poor have been left on the margins or omitted from negotiations completely, sometimes leaving them homeless. My own research (Abbott 1993, Abbott *et al.* 1998a, b) has led me to at least two disturbing examples, the demolition of Sanjay Gandhi Nagar (Nariman Point) as a result of pressure from the Colaba Residents Association (a prime site—see discussion below), and Subashnagar (Wadala) where in 1993 slum dwellers who had lived there for over 20 years were violently and forcibly evicted. In all these examples, those who have fared worse are lower-caste groups and Muslims who have often not even been consulted.

Property developers have also used environmental issues to make profits for themselves. For example, the long-standing attempts to improve Dharavi exemplifies how the rich have quickly found loopholes and ways to gain access to land meant to rehouse the poor (in spite of urban land ceiling limits) (Sanyal 2001). Also, as Mumbai has expanded, the easy access to its internal railway and its central location has made chawl land highly sought after. The investment potential of these properties has generated much conflict between the people who live there and those arguing for the compulsory purchase and sale of the chawls. An ongoing example is that of the rigorous attempts to push out the Koli (the original fishing tribe of Mumbai) who occupy the prime site of Coloba (Emmel and Soussan 2001), which overlooks the historic tourist hotspots of the Gateway of India and the ultra luxurious heritage Taj Hotel, one of the sites of the horrific terrorist attack in November 2008. The main point, therefore, is that whilst the rich and poor share space and can be mutually dependent, they often have conflicting interests as represented in a multitude of agendas.

Returning to the key question

To return to the key question identified in the introduction: I have thus far suggested the types of issues that give rise to contested environmental agendas. I have also hinted that it would appear that ultimately the powerful have the major say in defining public need and shaping the environmental agenda. A current example of this is the conflict over the brown agenda of the immediate needs of people who have set up illegal slums and the greens who want to preserve the scarce green space of the Sanjay Gandhi National Park (the Park) in Mumbai. This section relies heavily on Zerah (2007).

The Park is the only remaining large undeveloped area in the north of Mumbai. It is home to many wild animals including leopards, and is rich in flora and other fauna. It is also a historical site with over a hundred Buddhist caves. It has two lakes, which supply some of Mumbai's water. The Park has been protected throughout colonial times and post-independence through various Forestry and Wildlife Protection Acts. It is located between Mumbai and the peri-urban suburb of Thane, which has witnessed a massive increase in population in recent years. The result is that the Park is wedged between two giant urban masses.

Many migrants began to settle on the edges, thus encroaching on forest land. They sought income opportunities through tree felling, poaching, and even starting up stone quarrying 'factories'. As more arrived, the Bombay Environmentalist Action Group (BEAG) took out a public interest litigation case against the State Government of Maharashtra (Zerah 2007). The gist of the litigation was that the squatters represented an environmental threat to the green spaces and to wildlife preservation.

After a long drawn out process stretching from 1995 to 2003 (dotted periodically by pressure from increasing leopard attacks on the squatters), the final court judgement ordered rehabilitation of 33,000 people who could 'prove residency' prior to 1995, and removal of an estimated illegal 20,000 settlers accompanied by the demolition of the slum: 30,000 people were 'rehabilitated' far away in a village in Kalyan, miles from their work and social links. There were several protests and some NGOs assisted the squatters with representation, especially some 2,500 tribal groups who had inhabited the forest for generations. These groups were, however, also ordered to relocate in spite of the fact their lives had been entwined with forest preservation for centuries (Zerah 2007).

This is a difficult case as there is clearly a need to consider both the green and brown agendas. Many Indian activists and academics argue that the judiciary rulings have legitimized demolition of slums and reinforced social prejudices against the poor. They have also not given a fair hearing to the brown agenda of the poor or an equal voice to all stakeholders and the human suffering that went with the decisions, and not taken to task city planners who had allowed the situation to develop in the first place. They also argue that the judiciary favoured the rich as they focused on slum housing rather than the numbers of larger bungalows that had sprung up as permanent structures within the Park. Yet a green agenda favouring the preservation of the only national park in Mumbai cannot be dismissed, certainly when the preservation of Indian wildlife and heritage sites are of world interest.

Conclusion

Negotiating environmental protection is a complex story that reaches beyond what is immediately visible. It is a story woven around history, social settings, power, and powerlessness within intricate webs of multiple private agendas and privilege. Within this contest the most vulnerable, more often than not, will not be heard and a green agenda will win over a brown agenda.

A point of hope may be to understand that people are never passive recipients and contest will always continue. For instance, in recent years groups have begun to work collectively in negotiating brown agendas for themselves and have netted some positive results. For example, a slum called Shantinagar (Western Extended Suburbs) survived by resistance through collective community organization (Emmel and Soussan 2001). Another group negotiated rehousing through alliances with SPARC (an NGO), Mahila Milan (a women's organization), and the National Slum Dwellers association, and forced the Railway and Municipal Authorities to resettle railway bridge dwellers. In this, slum dwellers of Mumbai have received support from film-makers (such as Anand Patwardhan and Mira Nair) and writers (Arundathi Roy) hoping to reinforce the image of slums as places of living and livelihoods, and not merely the visible symbols of environmental degradation.

■ SUMMARY

- This chapter argues that slums are the most visible sign of urban decay and poverty and each slum locality has its own socio-economic and political history.

- In some cities such as Mumbai, the poor who live in slums share spaces with the rich who, in turn, live in luxurious housing.

- These shared spaces give rise to several contesting environmental agendas, both (i) between various groups of slum dwellers who negotiate for their own interests, and (ii) between the rich who want green and pleasant spaces and the slum dwellers who want housing.

- The powerful voices fighting for a green agenda often suppress the needy voices of a brown agenda.

■ REFERENCES

Abbott, D. (1993) Women's home-based income generation as a strategy towards poverty survival: the 'khannawalli' (mealmaking) activity of Bombay. Ph.D. thesis, The Open University.

Abbott, D., O'Hare, G. and Burke, M. (1998a) A review of slum housing policies in Mumbai. *Cities* 14(4): 269–283.

Abbott, D., Deshpande, S. and O'Hare, G. (1998b) Socio-economic survey of 16 slums in Mumbai. Carried out in colloboration between The University of Derby, Geographical Sciences, UK and The University of Mumbai, Dept of Economics India. Findings published in various articles by each of the authors, for example, Abbott *et al.* (1998a).

Beall, J. (2002) Water supply and sanitation for sustainable cities. In J. Beall, B. Crow, S. Simon and G. Wilson (eds), *International Development: Challenges for a World in Transition: Sustainability Theme*, pp. 92–140. Milton Keynes, Open University Press.

Burnett-Hurst, A. (1925) *Labour and Housing in Bombay: A Study in the Economic Conditions of the Wage-Earning Classes in Bombay.* London: P. S. King and Sons Ltd.

Davies, M. (2006) *Planet of Slums.* London: Verso, pp. 2–11.

Emmel, N. and Soussan, G. (2001) Interpreting environmental degradation and development in the slums of Mumbai, India. *Land Degradation and Development* 12: 277–283.

Government of India (2001) *Census of India.* New Delhi: Ministry of Home Affairs, Registrar General and Census Commissioner. Available from http://www.censusindia.net/.

Sanyal, B. (2001) Institutional pluralism and housing delivery: a case of unforeseen conflicts in Mumbai, India. *World Development* 29(12): 2043–2057.

Shellenberger, M. and Nordhaus, T. (2009) *Break Through: Why we Can't Leave Saving the Planet to Environmentalists.* Mariner Books.

Upadyay, S. (1990) Cotton mill workers in Bombay: conditions of work and life 1875–1918. *Economic and Political Weekly of India* 28 July: 87–99.

Zachariah, K. (1968) *Migrants in Greater Bombay.* Mumbai: Demographic and Research Centre Archives.

Zerah, M-H. (2007) 'Conflict Between Green Space Preservation and Housing Needs: The Case of the Sanjay Gandhi National Park in Mumbai', *Cities*, vol. 24, no. 2, pp.122–32.

■ FURTHER READING

Burra, S. (2005) Towards a pro-poor framework for slum upgrading in Mumbai, India. *Environment and Urbanization* 17(1): 67–88.

Dupont, V. (2007) Conflicting stakes and governance in the peripheries of large Indian metropolises—an introduction. *Cities* 24(2): 89–94.

Patel, S. (1990) Street children, hotel boys and children of pavement dwellers and construction workers in Bombay—how they meet their daily needs. *Environment and Urbanization* 2(2): 9–26.

Kraas, F. (2007) Megacities and global change: key priorities. *Geographical Journal* 173(1): 79–82.

■ USEFUL WEBSITES

http://www.mcgm.gov.in/ This is the official Website of Municipal Corporation of Greater Mumbai, giving information about the city, statistics, directory, downloadable forms, and feedback.

http://www.homeless-international.org/ Homeless International—India: SPARC . This is the website of an activist NGO called SPARC working for the rights of slumdwellers in Mumbai.

http://www.archidev.org/ The Asia Link of this website offers several archived articles resulting from joint research projects undertaken by Cambridge University, UK, Rizvi College of Architecture, Mumbai, India and the Centre for Environment Planning and Technology, Ahmedabad, India.

http://www.emagazine.com/ This is an online magazine covering a multitude of green topics.

Section B review

Section B has provided a fascinating insight into the diversity of perspectives on environment, development, and sustainability, while also noting the interconnections of the biophysical changes induced by human activity. It picks up directly the first premise of Chapter 1, that issues of environment and development are inextricably linked in chapters on China (2), Uganda (3), Ethiopia and Niger (4), Cuba (6) and Mumbai (9). While less obvious in other chapters of the section, the links are present in these also if we take the broad view of development, espoused in Chapter 1, that it is an issue for all countries of the world.

What also comes through from Section B is that, while the relationship between environment, development, and sustainability is a global concern, its effects are experienced very differently depending on locality. This leads to different foci in different places and different priorities for action.

Thus, where we live in the world certainly has a major influence in terms of framing our perspective on how we define and view the issues. However, even within locations—and hence within individual chapters in this section as well as between them—there are also different perspectives depending on socio-economic circumstance. Socially, places are not homogenous.

Kelly Gallagher illustrates, in the context of China (Chapter 2), an economic view of environment, development, and sustainability. In order to meet the developmental material needs of its huge population, China's economy has to grow, and fuelling that growth requires energy. Although, as she points out, the overall size of China's economy is already very large, there is a huge and growing economic inequality between people in China. Overall growth, which eventually benefits all to some extent, is seen as a way of heading off social unrest. In keeping with this essentially economic analysis, Gallagher also points out the high costs associated with ensuing environmental damage. These include inefficient use of resources, human health costs from polluted air and water, and the projected costs of the climate change-induced impacts of rising sea levels on coastal cities. Although the last is a global phenomenon, China is increasingly contributing to it through the sheer scale of its energy use and concomitant carbon emissions. On the plus side, China is fully aware of these costs and in some respects is an exemplar in mitigating them. It needs, however, technological investment to do more, which requires help from richer countries.

Economic development and its environmental costs is also the theme of Chapter 3 about Uganda by Margaret Mangheni, Mulondo Ssenkaali, and Fred Onyai. This country is small on most measures—population, overall size of the economy, and the annual per capita income. Schemes are promoted for growing the economy, such as commercial banana growing and the palm oil plantation project described in the chapter. While noting broad issues, such as biodiversity loss when forests are destroyed, the economics of the associated environmental costs are put more in terms of local livelihoods which can be lost, whether or not there are broader gains to the national economy. Also, there are potential social costs through these developments and the section on commercial bananas singles out negative consequences for gender equality. However, we should be careful not to romanticize local livelihoods and notions of living in harmony with the environment. The section on the Ssese islands illustrates this in the case of unsustainable fishing practices. When you are very poor, the welfare of future generations is rather less of a priority than earning a livelihood today. Charcoal production from the rainforests on the Ssese islands is also an issue, and we see here how local people have seized an economic opportunity that has literally burnt out to a large extent on mainland Uganda.

In case we start to think that it's all about the environmental costs associated with economic development, Kerstin Danert (Chapter 4) reminds us in the context of another two poor countries—Niger and Ethiopia—of the very positive human need for environmental resources in order to live. This is the sustainable provision of water, which is necessary for all life forms, not just humans. Rather than its use to the economy, Chapter 4 presents a view of environment in public health terms, of having safe drinking water and being able to irrigate crops grown for food. Danert also draws our attention to the robustness and effectiveness dimensions of sustainability introduced in the book's second premise in Chapter 1. Water supply, whether for drinking or irrigation of crops, needs to be robust to shocks (such as unexpected droughts) and effective in the sense of providing sufficient quantity of the right quality. How to nurture these dimensions, however, is contested. Are they to be nurtured through international donors providing 'improved' or 'modern' technologies, or through encouraging the efforts of local people and their communities themselves to improve their water supply incrementally over time? The former provide a kind of instant fix and may help meet international targets which aren't far away, such as the **Millennium Development Goals** which are to be achieved by 2015, but they may not be capable of being sustained for a variety of reasons. The latter approach is not going to result in fixes but may be the better way in the long term.

The incremental approach to sustainable water resources advocated by Danert rests on another idea—that local people do hold valid knowledge with respect to environmental improvement, and it is on this that we must build. The point is interestingly taken up by Petr Jehlička (Chapter 5) in the context of Eastern European countries and the environmental aspects of their accession to the European Union (EU). Thus, Jehlička points out that waste management regimes based on minimization, reuse, and recycling have long been prevalent in Eastern Europe, while EU policies are enforcing less environmentally friendly practices of incineration and landfill. Also practices of food self-provisioning (as witnessed by, for example, popularity of allotments) are far more widespread than in Western Europe.

Jehlička suggests that Eastern European approaches to waste management and food self-provisioning can together contribute to sustainable development under the banner of 'sustainable consumption'. The antecedents of sustainable consumption can be found within the thrift culture of the former Soviet bloc of which these countries were part, and of the desire to achieve a degree of freedom from central state control through, for example, self-provisioning. They contrast, therefore, with the Western European economic model (and that of the United States—see Chapter 8) that requires increasing consumption through 'green consumerism', and technological solutions to the mess created.

Cuba is an existing example of a centrally planned economy where austerity has been heightened because of the ongoing United States blockade and the collapse of the former Soviet Union which formed much of its export market. James Warren (Chapter 6) suggests, however, that these particular circumstances have contributed directly to a more sustainable transport system which is predicated on low car ownership and high (and cheap) public transport usage. There are issues over the investment required to renew the public transport system, but Warren argues that there is much here for the West to learn, given that private transport is such an energy guzzler and polluter.

Returning for the moment to the other existing centrally planned economy which features in this book, China, we can note that unlike Cuba it is expanding dramatically in economic terms. It is also a much larger country. In China, Gallagher argues that the absence and unacceptability of environmental groups which are allowed to criticize the state is a constraint on the formulation of more environmentally friendly policies. This is less the case in a country like Australia, where, according to Nicky Ison (Chapter 7), environmental activism is alive and well. Ison clearly believes that confrontation with power-holders is necessary through non-violent direct action, because of the scale of environmental problems and in the name of the overall well-being of the planet. She also firmly believes in demonstrating a kind of alternative, egalitarian, and open democracy through individual and collective lifestyles and in the way environmental groups organize.

The United States is the energy guzzler par excellence, and there too we find significant environmental activism. Ben Crow notes that such activism has a history which, in the United States, stems from eighteenth- and nineteenth-century notions of wild, untamed nature and frontier, notions which have helped construct the identity of being American. Thus, many environmentalists in the United States focus on reverence for, and hence the need to protect, wilderness. Historically this has resulted in the great national parks of the United States. The problem with this view of environment, Crow argues, is that it ignores the social dynamics of environmental change in the US the siting of toxic waste dumps next to poor and ethnic minority communities, for example. It also militates against forming alliances between the environmental movement and other groups such as trade unions, and hence opportunities for levering greater influence over policy are lost.

The American view of environment as protection of wilderness accords with an aspect of the final chapter in this section (9), where Dina Abbott contrasts the middle and upper class concerns in the mega-city of Mumbai of conserving open (and wild) space with the critical environmental health concerns of the slum dwellers. Both concerns have their activists, and, while the former have economic and political clout, the latter are certainly not passive in negotiating what Abbott terms brown agendas.

We see from this chapter, therefore, and several other chapters in the section, more than one sustainability in the different perspectives, as in the book's second premise (Chapter 1).

Although there are many nuances, broadly, we might divide them into two groups: those who emphasize sustainability of the biophysical environment and those who emphasize sustainability of people's livelihoods (whether considered economically or in social and environmental health terms).

We end this section review by briefly introducing the book's fourth premise, as introduced in Chapter 1. This concerns the dual nature of difference: as inequality and power, and as a resource for learning. We have seen throughout the section the former negative connotation, both within and between places. How, then, might we use difference as a resource for learning? The first and last chapters of the section give us a clue in that they imply an interdependence—in Chapter 2 an interdependence between economic development and maintaining a healthy environment; in Chapter 9 an interdependence between rich and poor in Mumbai. Look also for signs of this interdependence in other chapters of the section. We refer to the notion again in the section reviews for Parts C and D and develop it further in the book's conclusion (Chapter 27).

SECTION C
Major themes in environment, development, and sustainability

10

Climate change: causes and consequences

Roger Blackmore

Introduction

> *There is no bigger problem than climate change. The threat is quite simple, it's a threat to our civilization.*

> *Professor Sir David King, UK Government Chief Scientific Adviser, 2000–2007, King (2004)*

This chapter sets out to answer two questions about climate change. Why is the Earth getting warmer? How significant are the changes going to be? Much of the data and evidence supporting the answers has been provided by the Intergovernmental Panel on Climate Change (IPCC), set up in 1988 by the United Nations Environment Programme and the World Meteorological Organization to improve understanding about global warming. The IPCC is an extraordinary example of international and interdisciplinary collaboration between scientists and other academics across the world. Their efforts have advanced significantly our understanding of how the earth's physical and biological systems, its atmosphere, oceans, land, ice, and the living world including ourselves, interact and influence each other. At the same time the information and evidence underpinning this understanding has been debated openly and vigorously in the media.

Why is the Earth getting warmer?

We now know that the Earth is getting warmer. Eleven of the last twelve years (1995–2006) are the warmest observed since instrumental records began in 1850, and every year brings new evidence of changes to climate and weather, and their environmental impacts. According to the latest report from the IPCC:

> *Warming of the climate system is unequivocal, as is now evident from observations of increases in global average air and ocean temperatures, widespread melting of snow and ice, and rising global mean sea level.*

> *IPCC (2007a: 4)*

The report makes clear that this 'unequivocal warming' is outside the normal range of climate variability observed in recent centuries. In other words, the climate change now being observed is not just part of a natural cycle; there must be a new factor affecting the Earth's climate.

To understand what this might be, consider what determines the temperature of the Earth. Almost all the energy reaching the Earth's surface comes from the Sun (a much smaller flow comes from the Earth's hot interior). The Earth's temperature is controlled by the balance between the energy arriving from the Sun and that radiated out to space from the Earth as heat. This balance is influenced by four important factors: the temperature of the Sun; periodic changes in the Earth's orbit which affect the distance from the Sun to the Earth; the nature of the Earth's atmosphere, and the amount of sunlight reflected away from the Earth's surface (and thus not available to warm it). The last factor varies considerably with the weather and the seasons—snow and clouds reflect strongly, oceans and vegetation weakly.

Over timescales of thousands to millions of years the temperature of the Sun and the Earth's orbital patterns have varied. However, over the timescale of current global warming, approximately the last hundred years, these factors have not changed enough to explain the measured increase in temperature. What has changed significantly has been the composition of the atmosphere, in particular the concentration of certain gases critical to the temperature balance, known as greenhouse gases.

The greater part of the Earth's atmosphere, when dry, is made up almost entirely of three gases: nitrogen, 78 per cent by volume; oxygen, 21 per cent; and an inert gas, argon, 0.9 per cent. These three gases do not interact significantly with heat radiation from the Sun or the Earth. If these were the only gases in our atmosphere the Earth would be in a deep freeze, its mean temperature would be approximately $-18\,^\circ$C. The two main naturally occurring greenhouse gases are carbon dioxide and water vapour; others include methane and nitrous oxide. They are normally only present in small amounts, but their impact is significant. They act by absorbing the heat radiated from the Earth, and distributing it within the atmosphere. Most of the heat escapes to space, but, as Figure 10.1 illustrates, some is sent back to the Earth's surface and to the lower layers of the atmosphere where it is effectively trapped, warming both. The overall effect, known as the natural greenhouse

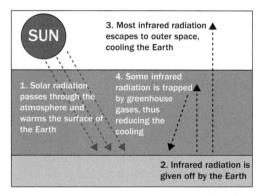

Figure 10.1 The greenhouse effect.

Source: HM Treasury (2006a).

effect, is to warm the Earth by about 30 $^{\circ}$C to its current mean value close to 15 $^{\circ}$C. Life depends on the warmth of the Sun being trapped in this way much as a greenhouse keeps warm on a sunny day; without it the Earth would not be habitable. (Note that the figures in this paragraph do not take account of reflected sunlight which, as noted above, is variable.)

The main reason for recent global warming can be described simply. (Sometimes a distinction is made between global warming, i.e. temperature rise alone, and global climate change, i.e. its wider impacts, but in this chapter the two are used interchangeably.) For several centuries, ever since the industrial revolution, Western societies have been extracting and burning fossil fuels such as coal, oil, and gas to provide energy for industry, domestic use, and transport, and as a result have been releasing carbon dioxide into the atmosphere. Today, most countries are heavily reliant on these sources for their energy. In addition, deforestation releases carbon dioxide, and the practices of modern agriculture release methane and nitrous oxide. This has set in motion a chain of events: the release of greenhouse gases into the atmosphere leads to higher concentrations in the atmosphere that, in turn, cause additional warming of the Earth. This new factor affecting the Earth's climate is called the anthropogenic or enhanced greenhouse effect, and it is caused by human economic activity.

Global emissions of greenhouse gases arising from economic activity have grown rapidly since pre-industrial times, due to a combination of growth in global population and growth in energy use per person. Between 1850 and 2000, global population rose from just over one billion to six billion and energy use per person rose sevenfold (Open University 2005). More people enjoyed a higher standard of living through increased consumption of goods and use of services, in spite of continuing poverty for many. In recent decades, the pace of greenhouse gas emissions has accelerated, increasing by 70 per cent between 1970 and 2004. After a short time lag the Earth's climate has responded by warming rapidly.

Figure 10.2a shows the contribution of the key greenhouse gases to total emissions in 2004, expressed as carbon dioxide equivalents, i.e. the amount of carbon dioxide that would have the equivalent effect. The most important greenhouse gas is clearly carbon dioxide. The main cause of carbon dioxide release is the burning of fossil fuel, which accounts for about three-quarters of carbon dioxide emissions. Land use changes, particularly deforestation, account for most of the rest with a small contribution from cement production. Methane (CH_4) is the next biggest contributor with 14 per cent, followed by nitrous oxide (N_2O) with 8 per cent. Agricultural practices are the main source of both of these gases. Figure 10.2b shows the contribution of economic sectors to global greenhouse gas emissions in 2004. This is the nub of the problem: almost all economic activities today contribute to emissions. Reducing these emissions presents many and diverse challenges.

The carbon stored in fossil fuels, which took millions of years to form, is being released back to the atmosphere in the space of a few hundred years. This rate of release far exceeds the capacity of the Earth's natural processes to remove it from the atmosphere by returning it to the oceans, soils, and plants. This is why atmospheric concentrations of all greenhouse gases have increased dramatically since the original UK industrial revolution in the eighteenth century.

Figure 10.3, which shows concentrations of carbon dioxide and other greenhouse gases in the atmosphere over the last 2000 years, illustrates this point. Until approximately 1750, concentrations of the three greenhouse gases remained roughly constant. Since

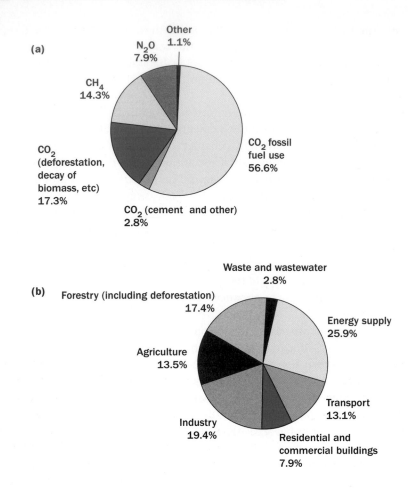

Figure 10.2 (a) and (b) Pie charts of global greenhouse gas emissions in 2004.
Source: IPCC (2007b).

then they have risen rapidly: carbon dioxide has risen by a third, methane has more than doubled, and nitrous oxide has increased by nearly 20 per cent.

How significant are the changes going to be?

[S]cientists are virtually screaming from the rooftops now. The debate is over! There is no longer any debate in the scientific community about this. But the political systems around the world have held this at arms length because it's an inconvenient truth, because they don't want to accept that it's a moral imperative.

Al Gore, An inconvenient truth. Gore (2006)

There are two reasons why scientists have been 'screaming from the rooftops' recently. The first is to do with timescales. Figure 10.3 clearly shows how much carbon dioxide

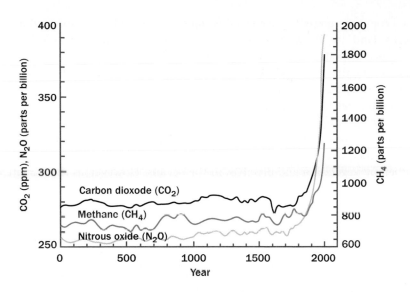

Figure 10.3 Concentrations of carbon dioxide and other greenhouse gases over the last 2000 years.
Source: IPCC (2007c).

and other greenhouse gases have increased in the last few hundred years, but how unusual is this on a longer timescale? Ice cores taken from sites on the Antarctic ice cap contain air trapped in bubbles that provide a record of the state of the atmosphere for the past 800,000 years. Analysis of this air (Lüthi *et al.* 2008) has shown that the Earth's climate over this period has swung eight times between long, cold glacial periods and warmer interglacial periods more like our present climate. Throughout this period, temperatures have varied from 6–8 °C colder than now to a few degrees warmer, while carbon dioxide concentrations in the atmosphere have fluctuated within a fairly narrow range of 180 ppm (parts per million) to 300 ppm. But the current level of carbon dioxide, 386 ppm (2008), is now much higher than this range and rising rapidly. The implication is clear. If current levels persist, let alone continue to rise, the Earth is likely to become much warmer than it has been for nearly a million years. While the Earth has experienced higher levels of greenhouse gases and warmer climates in its long geological history, we humans, who have only existed on the Earth for about 250,000 years, have not.

In stark contrast to these long timescales, the time horizons of many governments are short. Their primary focus is likely to be the next election or next crisis, not the next generation. It is hard for most governments and organizations to plan far ahead into the future even though energy supplies and infrastructure take decades to change, not years. Yet many of the greenhouse gases already added to the atmosphere will persist for the next 100 years, committing future generations to a warmer world and unknown climate shocks. So far, governments have found it hard to rise to Al Gore's moral challenge.

The second reason for growing concern comes from the computer models that climate scientists use to explore the interactions between the atmosphere and other parts of the Earth.

Increased computer power has allowed their models to include more interactions between the oceans, land, ice, the living world, and the atmosphere, leading to realistic outputs. Global circulation models are now able to model past and current climates accurately enough to demonstrate both that global warming is now occurring and that it cannot be caused by natural variability alone—it can only be explained by human activity. They are also warning that future global warming may be more damaging and more rapid than earlier models have suggested. Recent modelling outputs are starting to produce some scary results.

Two examples, one from modelling and the other from the real world illustrate this point. Currently, over half of the additional carbon dioxide added to the atmosphere is absorbed by vegetation and dissolves in the oceans. However, as the climate warms, warmer oceans hold less carbon dioxide, bacteria in warmer soils break down organic matter faster and release more greenhouse gases, and some tropical forests could die back as drought becomes more common. Results from recent models (Cox *et al.* 2000, Cox and Jones 2008) suggest that, as warming progresses, forests and oceans will lose some of their ability to take up carbon dioxide and in some cases may start to release it. In either case the result would be to amplify and accelerate the effects of global warming.

The second example is the effect on the climate of disappearing glaciers, snow, and ice in a warmer world. Snow and ice reflect most of the sun's energy back to space, keeping snow-covered areas cooler. When there is less snow, this cooling no longer happens, so in regions where snow and ice are in retreat, for example around the Arctic, warming is expected to be enhanced. Models show this by predicting a much greater rate of warming in and around the Arctic than elsewhere on the globe. What they have not predicted is the speed of the retreat of Arctic ice in the summer, or of glaciers on Greenland (Pfeffer *et al.* 2008). In the summers of both 2007 and 2008 the sudden decline of Arctic ice surprised and alarmed the people monitoring events (NSIDC 2008).

For scientists, as Al Gore pointed out, the debate about the reality of climate change is now over. The focus has moved on to a better understanding of its impacts, and providing more useful forecasts for governments and communities. A major focus is to communicate to leaders at all levels the significance of future changes and the need for prompt action. One approach is to predict more specific impacts expected at different levels of warming. For example, the UK Government Stern Review (HM Treasury 2006a) predicted that a 1 °C rise will mean the disappearance of small mountain glaciers and severe damage to coral reefs. A 2–3 °C increase will see significant changes in water availability and falling crop yields in many developing regions. With an increase of 4 °C, sea level rise will threaten major world cities including London, Shanghai, New York, Tokyo, and Hong Kong. These examples illustrate what might happen if adaptive measures are not taken, but it's not possible to predict with precision that a particular event will occur at a particular temperature. The effects of future climate change are still difficult to predict accurately because they are unlikely to act in isolation and will add to existing pressures on vulnerable human and natural systems.

In spite of these caveats, the general trends should be clear. If global temperatures rise to 2°C or more above pre-industrial levels—and the Earth is already half a degree warmer now—a series of major, damaging impacts are predicted to occur.

A rising number of people are threatened by hunger and from regional water shortages, and at risk from a rising intensity of storms, droughts, floods, and heatwaves. In addition, irreversible processes may begin to destroy the Amazonian rainforest and melt the Greenland ice sheet (which would add 7 metres to sea levels), and a significant proportion of all species may become extinct.

Paths to action

The goal of climate change policy makers has always been to prevent this situation occurring. This is stated in Article 2 of the *United Nations Framework Convention on Climate Change* (UNEP/WMO 1992: 4):

The ultimate objective ... is to achieve ... stabilisation of greenhouse gas concentrations in the atmosphere at a level that would prevent dangerous anthropogenic interference with the climate system. Such a level should be achieved within a time frame sufficient to allow ecosystems to adapt naturally to climate change, to ensure that food production is not threatened and to enable economic development to proceed in a sustainable manner.

Many would consider that 'dangerous interference' is already occurring and that the examples just described demonstrate that the rise in global temperature needs to keep below 2 °C to avoid the worst consequences of climate change. To achieve this goal, models now indicate that concentrations of all greenhouse gases in the atmosphere need to be held back to the equivalent of 450 ppm carbon dioxide. This would require global emissions to fall eventually by 80 per cent or even 90 per cent of current levels – a much tougher target than had been thought only a few years ago, and one that requires action be taken as soon as possible to have any prospect of success. In the words of the head of climate change for the UK government 'If action is delayed or is slow, then there is a significant risk of much larger increases in temperature' (Pope 2008).

A further issue concerns the unequal distribution of the impacts of climate change. The wealthy can afford the measures needed to adapt more than the poor. The subsistence farmers and fishermen of the Ganges or Nile deltas (and many others) are far more vulnerable to climate hazards than Londoners protected by the Thames barrier, and have fewer financial and other resources to help them recover from rising sea levels, floods or storms. In general, developing countries are much more vulnerable than developed, and the weakest economies are the most vulnerable. Some regions are particularly susceptible due to their geography, for example small islands and the large delta regions of Africa and Asia, while in others, such as Arctic nations and dry areas in low latitudes, the early effects of climate change are likely to be particularly acute.

The unequal distribution of impacts is of importance for another reason. It contrasts the fortunes of developed countries that have been responsible for most greenhouse gas emissions until recently, with developing countries that have not but are suffering disproportionately. Inequity has dogged global negotiations to reduce greenhouse gas emissions. Developing countries want to see developed countries acting first, whereas developed countries are fearful that if countries such as China follow their path of development, then greenhouse gas emissions will soar. To find some way round this stand off – blame is not a likely foundation of a constructive agreement – Professor Stephen Pacala (a US climate scientist) turned his attention to the behaviour of individuals rather than nations. He did this by estimating the carbon dioxide emissions of everyone in the world ranked by their income, a task he described as 'not particularly easy'. What he found was a stark contrast between the carbon dioxide emissions of the poor and the rich. The contribution from the poorest half of humanity is less than 7 per cent of the total.

> *The 3 billion poorest people … emit essentially nothing. The take-home message here is that you could increase the emissions of all those people by putting diesel generators or anything you wanted into their lives and it would not materially affect anything I'm going to say [about stabilising carbon dioxide to 450 ppm]. In other words, the development of the desperately poor is not in conflict with solving the climate problem, which is a problem of the very rich. … In contrast, the rich are really spectacular emitters.*
>
> Pacala (2007)

The rich he described as the top 500 million who earn somewhat more than the average US citizen, and are found in all countries, rich and poor. These 7 per cent of humanity emit half of all emissions, and bearing down on the high emitters is necessary to have any chance of reaching the targets. However, Pacala's approach changes the focus from countries and actions by 'them', to individuals and our own responsibilities. I suspect that some of the authors of this book, and I include myself, while not the richest individuals, are creating very high carbon footprints as we fly from conference to conference. Change has to begin at home.

Conclusion

When discussing climate change it is all too easy to be overwhelmed by the scale of the problems it poses and the inadequacy of most political responses or the effect of individual action. To avoid the dangers and damaging impacts of climate change, there is clearly a need both to reduce emissions of greenhouse gases dramatically and to prepare for and adapt to the changes already in the pipeline. It is time, however, to move away from the narrative of doom and gloom to one of opportunities and solutions. Continents in the tropics and subtropics, for example Africa, South America, and southern Asia, are going to be hardest hit by climate change, yet they hold most of the planet's renewable resources in the form of energy from sunlight, rainfall, and fresh water, stores of biodiversity, and young people. Their populations mostly lack capital to develop these resources in sustainable ways, but they have the potential and the natural resources to be a major part of the solution to the global challenges of climate change and sustainable development.

Simple mechanisms such as the transfer of new, clean technology coupled with fair trade could transform the outlook. As Stern points out in his review of the economics of climate change (HM Treasury 2006b), acting now and committing spending now is going to be much less costly in the medium term than doing nothing, and there is still time to avoid the worst impacts of climate change.

■ SUMMARY

- The Earth's climate has warmed in recent decades due to changes in its atmosphere, specifically increases in carbon dioxide and other greenhouse gases.

- The cause is global emissions of greenhouse gases by societies using fossil fuels and industrial processes.

- The level of greenhouse gases now far exceeds that of the last 800,000 years and is increasing rapidly.

- Greenhouse gases persist in the atmosphere for centuries rather than years, already 'committing' the Earth to significant further warming.

- Governments usually have short time horizons while societies can take decades to change infra-structures.

- Climate models are signalling a growing risk that climate change may be more rapid and damaging than earlier thought.

- Climate impacts become more damaging with rising temperatures, affecting all societies and ecosystems, but are spread unequally. The most vulnerable people are the poor, who are not the cause of the problem.

■ REFERENCES

Cox, P. M., Betts, R. A., Jones, C. D., Spall, S. A. and Totterdell, I. J. (2000) Acceleration of global warming due to carbon cycle feedbacks in a coupled climate model. *Nature* 408: 184–187.

Cox, P. M. and Jones, C. (2008) Climate change: illuminating the modern dance of climate and CO_2. *Science* 321: 1642–1644.

Gore, A. (2006) *An Inconvenient Truth*. Paramount Classics, Paramount Pictures, USA.

HM Treasury (2006a) *Stern Review: The Economics of Climate Change, Part 1: Climate Change—Our Approach*. London: HM Treasury London. Available from http://www.hm-treasury.gov.uk/sternreview_index.htm, accessed 17 June 2009.

HM Treasury (2006b) *Stern Review: The Economics of Climate Change, Executive Summary*. London: HM Treasury. Available from http://www.hm-treasury.gov.uk/sternreview_index.htm, accessed 17 June 2009.

IPCC (2007a) Summary for policymakers. In Contribution of Working Group I to the Fourth Assessment Report of the Intergovernmental Panel on Climate Change, *Climate Change 2007: The Physical Science Basis*, p. 4. Cambridge: Cambridge University Press.

IPCC (2007b) Summary for policymakers. In Contribution of Working Group I to the Fourth Assessment Report of the Intergovernmental Panel on Climate Change, *Climate Change 2007: Synthesis Report of the Fourth Assessment Report of the Intergovernmental Panel on Climate Change*, p. 6. Cambridge: Cambridge University Press.

IPCC (2007c) Summary for policymakers. In Contribution of Working Group I to the Fourth Assessment Report of the Intergovernmental Panel on Climate Change, *Climate Change 2007: The Physical Science Basis*, FAQ 2.1. Cambridge: Cambridge University Press.

King, Professor Sir David (2004) Quoted in an interview with The Climate Group, The Office of Science and Technology, London. Available from http://www.theclimategroup.org/news_and_events/professor_sir_david_king/, accessed 20 October 2008.

Lüthi, D., Le Floch, M., Bereiter, B., Blunier, T., Barnola, J-M., Siegenthaler, U., Raynard, D., Jouzel, J., Fischer, H., Kawamure, K. and Stocker, T. F. (2008) High-resolution carbon dioxide concentration record 650,000—800,000 years before present. *Nature* 453: 379–382.

NSIDC (2 October 2008) Arctic sea ice down to second lowest extent; likely record low volume. Press release from the National Snow and Ice Data Centre, University of Colorado, USA. Available from http://nsidc.org/arctic/news/, accessed 20 October 2008.

Open University (2005) *Population and Energy Use*. Available from http://www.open.ac.uk/T206/longtour.htm, accesssed November 2005.

Pacala, S. (2007) Equitable solutions to greenhouse warming: on the distribution of wealth, emissions and responsibility within and between nations. Address to the 35th Conference of the International Institute for Applied Systems Analysis, 13–16 November 2007, Vienna. Available from http://www.iiasa.ac.at/Admin/INF/conf35/docs/programme.html?sb=3, accessed 20 October 2008.

Pfeffer, W. T., Harper, J. T. and O'Neel, S. (2008) Kinematic constraints on glacier contributions to 21st-century sea-level rise. *Science* 321: 1340–1343.

Pope, V. (2008) The Met Office's bleak forecast on climate change. *Guardian* 1 October.

UNEP/WMO (1992) *United Nations Framework Convention on Climate Change*. Geneva: United Nations Environment Programme and World Meteorological Organisation, Information Unit on Climate Change.

■ FURTHER READING

Henson, R. (2008) *The Rough Guide to Climate Change*, 2nd edn. London: Rough Guides.

Maslin, M. (2008) *Global Warming: A Very Short Introduction*, 2nd edn. Oxford: Oxford University Press.

Lynas, M. (2007) *Six Degrees: Our Future on a Hotter Planet*. London: Fourth Estate.

■ USEFUL WEBSITES

http://www.metoffice.gov.uk/climatechange/: this UK Meteorological Office Hadley Centre site summarizes key facts and effects from climate change and suggests what you can do to reduce your own impact. It gives a primer on climate science and shows the latest projections from its climate models.

http://www.ukcip.org.uk/: the UK-focused site, the UK Climate Impacts Programme (UKCIP) is a resource for individuals and organizations seeking to adapt to climate change. Good sections on climate science and UK climate change forecasts.

http://www.ec.gc.ca/: Environment Canada is one of my favourite national environment sites. It has good sections on climate change science, latest news on climate change and much more.

http://www.ipcc.ch/: the Intergovernmental Panel on Climate Change allows you to download or view any section of its latest report. If you are interested in climate change in a particular region look at the chapters in the WGII report. The site also has the latest news on the UN climate change negotiations.

11

Can ecosystems be managed sustainably?

Biodiversity and tropical forests

Michael Gillman

Introduction

Sustainable development, defined as development that meets the needs of the present without compromising the ability of future generations to meet their own needs (United Nations 1987), has been seen as the panacea for a range of environmental problems. However, it has become clear that the promise of sustainable development is extremely difficult to deliver. There is no easy win–win solution for populations of humans looking to exploit, even in a sensitive manner, increasing fractions of their environment.

This chapter explores a key aspect of sustainable development, namely **biodiversity** which for our purposes can be understood as the variation of plant and animal forms within an **ecosystem**. The chapter considers biodiversity with respect to the harvesting of components of tropical forests by humans. This will lead to the wider question of whether ecosystems, of which tropical forests are but one example, can be managed in a sustainable manner.

Why should humans be concerned with the sustainable maintenance of ecosystems such as tropical forests and their component parts? The first and most obvious answer is that these ecosystems are of direct benefit to humans, for example, in maintaining water cycles, preventing soil erosion, and providing resources such as timber or medicinal plants. Areas of woodland may provide habitats for pollinating insects vital to crop production. At a global level the plants in various ecosystems capture carbon dioxide from the atmosphere through photosynthesis and thereby contribute to the regulation of global climate. Collectively, these regional and global benefits to humans are seen as ecosystem services and have a high economic value. For example, the provision of pollination services by honey bees for crops in the United States in the year 2000 was estimated at US$14.6 billion (Greenleaf and Kremen 2006).

> Photosynthesis is the process whereby green plants in sunlight convert carbon dioxide and water into carbohydrates and oxygen.

The second reason to maintain ecosystems is for their intrinsic value as homes to a wide variety of organisms. There is no easy economic equivalence here, but the importance of conserving the planet's biodiversity is clearly something which motivates many humans.

Managing forests sustainably

Ecological theories of harvesting have a long history. Early work on fisheries in the 1950s (Ricker 1954) suggested simple models for optimum levels of harvesting resulting in the maximum sustainable yield, beyond which the fish stocks would go into decline (Figure 11.1). These studies focused on single species and demonstrated that certain levels of harvesting could be sustained as long as characteristics of the harvested population, such as growth rate and age-specific survival, were maintained at sufficient levels. Inevitably those models were difficult to reconcile with reality as populations did not behave in the required manner. Migration and complex interactions with predators and prey meant that new models had to be generated. Furthermore, the control of harvesting levels is more easily met in a computer programme than on a fishing fleet covering large areas of ocean. Terrestrial equivalents of the fisheries models were combined with economic considerations and applied to high-profile organisms such as African elephants for ivory (Barbier *et al.* 1990).

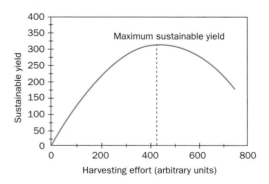

Figure 11.1 Single species model of sustainable harvesting. The sustainable yield of resource (which may be fish or trees) has a maximum value (maximum sustainable yield, MSY) with respect to increasing harvesting effort. Adapted from Ricker (1954).

A parallel exercise was undertaken in forest management with the aim of removing a fixed amount of timber without damaging the ecosystem within which the trees were growing. Again, these practices are much more straightforward in plantation systems where tree growth is easily monitored and the fate of other components of the ecosystem is a secondary consideration. Indeed, plantations with few associated species often replaced the original woodland resulting in a considerable local loss of biodiversity, for example, the replacement of deciduous woodland by coniferous woodland in lowland Britain. These practices, especially in areas of ancient woodland, are now being reversed in many parts of the world. Instead, the focus in both temperate and tropical systems is on management of the naturally regenerating woodland. The sustainable management of natural forests, with the promise of long-term returns from timber and maintenance of the ecosystem biodiversity and function, is now a widely held goal and seen as a win–win scenario for conservation and economics. Yet is this achievable?

An encouraging early form of natural forest management was the coppice cycle practised in woodlands of Western Europe over many hundreds of years, whereby young tree stems were repeatedly cut down to near-ground level leaving a stool from which new shoots emerged. Thus, under this management regime, trees such as hazel (*Corylus avellana*) or sweet chestnut (*Castanea sativa*) were cut at fixed intervals (a typical hazel harvest cycle is once every 10 years). Larger trees such as oaks would be left as standards which could then be harvested later for major construction projects. Different compartments of the woodland would be opened in subsequent years resulting in a mosaic of patches of varying ages. These patches would support a range of woodland species.

The maintenance or reinstatement of coppicing has been a successful conservation tool. For example, in the United Kingdom the regionally rare butterflies, pearl-bordered fritillary (*Boloria euphrosyne*) and heath fritillary (*Mellicta athalia*), thrive in coppiced areas, where their larval host-plants (such as dog violet, *Viola canina* and cow wheat, *Melampyrum pratense*) occur at high densities, i.e. high numbers per unit area. The heath fritillary (Figure 11.2) is a notable conservation success story which has made UK national headlines (*Independent* 22 June 2007). An important ecological principle relevant to these studies is that of high diversity associated with intermediate disturbance. Coppiced areas that are too open or too closed tend to have a lower species diversity compared with areas in coppice mid-cycle.

Similar approaches have been undertaken in tropical forests. The Periodic Block System of management in some of the lowland forest in Trinidad involves blocks of 150–300 hectares (ha) in which only trees meeting certain ecological criteria are harvested over a 2-year period. A set of environmentally sensitive practices is included, such as leaving large fruiting trees or those next to streams. After harvesting, the compartment is closed for 25–30 years. Studies of the component flora and fauna revealed that this management technique maintained higher population levels and higher richness of key groups such as epiphytes when compared to either plantations or continual cropping systems where there is no significant period of closure after harvesting (Clubbe and Jhilmit 2002; Wood and Gillman 1998) and that richness approached that of natural systems after 25 years (Table 11.1). Indeed, there was no evidence that species characteristic of undisturbed woodland were affected at all by the Periodic Block management, although there was an influx of species which thrive in open areas of

Figure 11.2 (a) Heath fritillary (*Mellicta athalia*) (b) coppiced woodland.

forest or at forest edges (referred to as gap species). In contrast, the continual crop-ping system had fewer true woodland species including epiphyte species. The higher number of undisturbed forest species in the early stage of the Periodic Block System may just be a chance effect of sampling or an increased opportunity for colonization by those species.

The example in Trinidad is on a relatively small-scale and works in part because it is tightly controlled. In larger operations in regions such as Amazonia and Borneo sustainable forestry may be more difficult, as we will see later.

There is now a substantial body of data demonstrating the effects of logging on differ-ent species and biodiversity in general. Largely those studies are consistent in identifying particular groups of species most at risk. However, there are some uncertainties, partly dependent on the type of logging being undertaken. A summary of the responses of squirrels to logging in Borneo, measured as changes in density, is given in Table 11.2 (Mei-jaard and Sheil 2008). Each density trend is supported by a different publication. So, for example, two out of three studies on the tropical squirrel *Ratufa affinis* indicate a decline

> Epiphytes are plants that live on other plants and are typical of tropical forest. They include orchids and bromeliads.

Table 11.1 Effect of the Periodic Block tropical forest management system on non-target species in Trinidad compared with continual cropping and primary forest (adapted from Clubbe and Jhilmit 2002). The primary forest received no management and was not subject to any disturbance from local people. Non-target species were sampled in eight 100 m² plots. Woodland species are those characteristic of undisturbed woodland and include epiphytes.

	Primary forest (no management)	Periodic Block System (end)	Periodic Block System (early)	Continual cropping system
Period since last logging	n/a	25	6	4
Total species richness	137	148	175	166
Total gap species	14	27	44	55
Total woodland species	123	121	131	111
Total epiphyte species	16	15	16	9

or loss from logged areas. Overall, on these twelve species, ten studies showed an increase in density after logging whilst fifteen showed a decrease.

A major concern is that, at least amongst mammals, species most at risk from logging may be those that evolved earlier—ancestral species (Meijaard and Sheil 2008). Studies in Borneo recognized that logging-intolerant species were those that started to evolve during the Miocene (up to 23 million years ago) and were characterized by little morphological variation, rarity on small islands, and narrow ecological niches with strictly fruit feeding, insect feeding, or carnivorous habits. It is perhaps not surprising that species with specialized feeding habits and low dispersal abilities should do less well than generalists in managed forest.

Table 11.2 The response of squirrel species to logging in Borneo (Meijaard and Sheil 2008).

Species	Ecological stratification	Change in numbers after logging
Ratufa affinis	Canopy	Two out of three populations decreased
Ratufa bicolor	Canopy	Population decreased
Callosciurus notatus	Middle of canopy	Two populations increased and two decreased
Callosciurus prevostii	Canopy	Population increased
Callosciurus caniceps	Middle of canopy	Population increased
Callosciurus nigrovittatus	Middle of canopy	No change
Rhinosciurus laticaudatus	Ground	Two populations decreased
Lariscus hosei	Ground	Population decreased
Lariscus insignis	Ground	Two populations decreased
Sundasciurus hippurus	Mixed	Three populations decreased and one population increased
Sundasciurus lowii	Mixed	Two populations increased and one decreased
Sundasciurus tenuis	Mixed	Two populations increased and one decreased

However, the linkage of those traits to early evolutionary origins does worry conservationists who wish to see the preservation of the earlier stems of the Earth's evolutionary tree.

Based on these and similar data sets, authors have made recommendations concerning the control of logging to ensure that impacts on non-target species are minimal. These recommendations are summarized below, based on Clubbe and Jhilmit (2002) and Meijaard and Sheil (2008):

- Improved legislation and law enforcement and increased accountability.

- The regulation of hunting in logged or recently logged areas where species are more vulnerable (e.g. more visible).

- Reduction of incidental damage to forests during harvesting.

- Preservation of canopy (the trees occupying the highest level of the forest) and mid-canopy fruiting trees to benefit fruit-feeding species and maintain seedling recruitment.

- Better protection of interior forest conditions. For example, trails should be planned carefully and constructed in a way that minimizes canopy damage.

- Leaving of dead or partially dead trees standing or intact on the ground.

- Preventing streams from silting up.

- Maintenance of interior forest connectivity and connections between forest patches.

- Implementation of adequate recovery periods.

With reference to the last point, we have seen that the Periodic Block System in Trinidad recommends a 25–30 year cycle. These forests are dominated by a few species such as *Mora excelsa* so that the cutting cycle is (partly) tuned to the recovery rate of those species. However, not all species grow at the same rate so if the cycle is too short the management system will start to shift the species composition in favour of the faster-growing trees. Furthermore, there is evidence of declining economic returns from cutting cycles of 25 years. A study in Bolivia (Dauber *et al.* 2005) has demonstrated that that a cutting cycle of this duration results in reduced returns of between 72–96 per cent in different eco-regions. Losses of this magnitude mean that there may be large areas of Amazonia and other tropical regions where natural forest management is only economically viable as a one-off activity, with potentially detrimental effects on the forest ecosystem.

Non-timber forest products: an economic and ecological alternative to timber?

An alternative to timber production in forests is to remove smaller components of the system. These may include products for minor construction such as palm leaves for thatch, medicinal plants, or products from animals such as honey. These products, collectively referred to as non-timber forest products (NTFP), could also include animals (e.g. mammals, birds, fish, lizards) consumed for meat, although this is generally considered separately as bush-meat. However, the same principles of sustainable utilization will apply to bush-meat and NTFP.

One clear difference between timber and non-timber utilization of forests is the economic scale of activity. Whilst individuals may collect small amounts of a single palm for making a basket, the large-scale production of timber requires heavy machinery and substantial forest areas for economic returns. Traditional communities have used timber for constructing their own houses or transport such as dug-out canoes, but this would rarely have been undertaken (or, indeed, have been feasible) for larger economic return. A corollary of the financial costs and benefits between timber and non-timber products is that the former tend to be undertaken by major companies whereas the latter are the domain of individuals, family, or small community units.

Acquisition of NTFPs may be for personal use, to support small-scale industries such as weaving or carving and sometimes for export for pharmaceutical and food production. Whilst this might suggest that NTFPs are likely to be a minor drain of the resources of forest systems, this is not necessarily the case. It is possible that a well-managed timber concession may have less long-term impact than a sustained period of collecting for NTFPs by a forest community. Fortunately there are now a number of studies across the tropical forests of the world to allow us to determine the effect of NTFP utilization. Examples of NTFP utilization in two montane forests in East Africa (Ndangalasi *et al.* 2007) are shown in Table 11.3.

Ndangalasi *et al.* (2007) concluded that levels of harvesting in both forests are higher than can be sustained, i.e., that continued use at those levels would lead to detrimental effects on the populations of plants themselves and potentially of the ecosystems within which they reside. A major difficulty is that setting levels of utilization is seen as meddling by outsiders whilst enforcement of harvesting levels would be unworkable.

Other studies of NTFP utilization have considered the economic returns. Shone and Caviglia-Harris (2006) addressed the environmental economics of NTFPs in western Amazonia. They considered whether the welfare of forest inhabitants could be maintained or improved and tropical deforestation reduced by utilization of NTFPs. Using a substantial data set and economic statistical analyses they showed that whilst agro-forestry practices which involved reduced deforestation and replanting did result in higher diversity, there was no link between the reduction of poverty and reduced deforestation. In essence it seems that deforestation may be more profitable for people. Certainly, there is far less

Table 11.3 Use of non-timber forest products in two tropical forests in East Africa

	Number of species used
Medicinal plants	57
Building poles	50
Wood fuel	35
Edible fruits and vegetables	21
Tool handles	9
Weaving and basketry	8
Thatch material	7

reliance on the forest for livelihoods now. The opportunities for raising chickens or cattle are much greater and provide a more reliable source of protein than bush-meat.

An argument is that, as human population size increases, it is no longer possible for small communities to persist in a sustainable existence with their forest environment. Even simple tools such as machetes mean that communities can quickly denude local areas of prized NTFPs. The traditional interactions of human communities with their forest environments may have been sustainable in the past only because human population size and activity were limited by the same natural processes of disease and nutrient limitation as the populations of other organisms in the forest.

It is clear that if tropical forests are to survive then local communities, national legislative bodies, and international agencies have to take active measures to protect them and not rely on their value as a timber or non-timber source. Whilst international concern over deforestation's contribution to climate change (see Chapter 10) may result in some incentives for conservation, there is a concern that such measures will not prioritize areas of highest biodiversity. The debate over the biological value of tropical forests has never suffered from lack of data. The same is now true over the debate on sustainable management.

■ SUMMARY

- Maintaining ecosystems can make economic sense, while many also value ecosystems and their biodiversity intrinsically.

- The Periodic Block System of tropical forest management with respect to timber production in some parts of the world offers possibilities for maintaining biodiversity and regeneration. Some research suggests, however, that it might not be economically viable because of the long regeneration periods that need to be allowed for recovering blocks.

- It is often argued that non-timber forest products (NTFPs) offer an ecologically sound economic alternative to timber. This argument is only valid, however, when it applies to low human population densities of forest dwellers. It does not necessarily alleviate poverty for forest dwellers as is also often claimed.

■ REFERENCES

Barbier, E. B., Burgess, J. C., Swanson, T. M. and Pearce, D. W. (1990) *Elephants, Economics and Ivory*. London: Earthscan.

Clubbe, C. and Jhilmit, S. (2002) Integrating forestry and biodiversity conservation in tropical forests in Trinidad. In M. Maunder, C. Clubbe, C. Hankamer and M. Groves (eds) *Plant Conservation in the Tropics*, pp. 185–206. Kew: The Royal Botanic Gardens.

Dauber, E., Fredericksen, T. S. and Pena, M. (2005) Sustainability of timber harvesting in Bolivian tropical forests. *Forest Ecology and Management* 214: 294–304.

Greenleaf, S. S. and Kremen, C. (2006) Wild bees enhance honey bees' pollination of hybrid sunflower. *Proceedings of the National Academy of Sciences, USA* 103: 13890–13895.

Independent (2007) *Heath fritillary returns as tree-cutting brings light back to the woods*. 22 June. Available from http://www.independent.co.uk/environment/nature/heath-fritillary-returns-as-treecutting-brings-light-back-to-the-woods-454161.html.

Meijaard, E. and Sheil, D. (2008) The persistence and conservation of Borneo's mammals in lowland rain forests managed for timber: observations, overviews and opportunities. *Ecol Res* 23: 21–34.

Ndangalasi, H. J., Bitarih, R. and Dovie, D.B.K. (2007) Harvesting of non-timber forest products and implications for conservation in two montane forests of East Africa. *Biological Conservation* 134: 242–250.

Ricker, W. E. (1954) Stock and recruitment. *J Fisheries Res Board Canada* 11: 559–623.

Shone, B. M. and Caviglia-Harris, J. L. (2006) Quantifying and comparing the value of non-timber forest products in the Amazon. *Ecological Economics* 58: 249–267.

United Nations (1987) *Report of the World Commission on Environment and Development. General Assembly Resolution 42/187.* New York: United Nations.

Wood, B. and Gillman, M. P. (1998) The effects of disturbance on forest butterflies using two methods of sampling in Trinidad. *Biodiversity and Conservation* 7: 597–616.

12

Conservation and development in Scotland and Uganda

Adeline Muheebwa and Roger Wheater

Introduction

Balancing conservation against economic pressures is a long-standing concern, with implications for both present and future generations. It is as important as ever in developed and developing countries alike.

Conservation has three main dimensions:

1. *Social/cultural heritage*: this refers to the values and traditions that people associate with the 'natural' environment. For example, Crow (Chapter 8 this volume) discusses reverence for wilderness as a driver of environmental concerns in the United States.

2. *The physical environment*: this concerns the conservation of species (both plant and animal) and their physical habitats of land, air, and water. It also concerns the interactions between the individual elements so that we can speak of conservation of ecosystems. It links, therefore, to the more recent concept of **biodiversity** (Chapter 11 this volume). In the section below, the physical environment in Scotland's national parks is referred to as the 'natural heritage', thereby taking into account its enduring characteristics over time.

3. *Local livelihoods*: rural people in developing countries often depend on natural resources directly for their livelihoods. They have an economic interest in conservation, therefore, in the sense that they wish to conserve what they exploit for its future as well as its present value.

These dimensions overlap. For example, the spiritual association with a physical site or landscape in our consciousness leads to demands for its conservation. The spiritual association often has a material foundation, in that the site's physical resources form (or have historically formed) the basis of local livelihoods.

This chapter explores the conservation challenge in two contrasting countries: Scotland (northern UK) and Uganda (east Africa). The above dimensions make it clear that this is not just a technical exercise in either country, and the two cases illustrate the multiple stakeholders who are, or should be, involved in conservation decisions.

Scotland: achieving a delicate balance between conservation, economic interests, and recreation through National Parks

Scotland is small in size and population, and is governed as part of the UK but with recently devolved powers in many areas. The population is concentrated in a few cities and the central belt of the country. The sparsely populated mountains, hills, rivers, and lochs (lakes) of the rest provide numerous recreational activities, including mountain climbing, hiking, canoeing, and sailing, as well as the traditional field sports of fishing, shooting, and deer stalking. A number of designations are aimed at conserving native species and areas of outstanding scenic beauty.

Although raised as a possibility in the 1930s, National Parks in Scotland did not receive the necessary legal framework until the year 2000 with the passing of the National Parks (Scotland) Act and its requirement to achieve a balance between conservation and economic activity. The Act requires that, in establishing parks, the area is of outstanding national importance either because of its natural heritage or a combination of natural and cultural heritage. The area must also have a distinctive character and a coherent identity.

National Parks in Scotland have four aims which together relate to the three dimensions of conservation in the chapter introduction. These are to:

1. Conserve and enhance the natural and cultural heritage (conservation of the physical environment and the social/cultural heritage).

2. Promote sustainable use of natural resources (conserve the physical environment and local livelihoods).

3. Promote understanding and enjoyment of the special qualities of the area by the public (conserve the social/cultural heritage).

4. Promote sustainable economic and social development of the area's communities (conserve local livelihoods).

All matters relating to the creation, development, review, or major changes in approved plans of such parks must be subjected to public consultation across a range of stakeholders. If there is fundamental disagreement a local enquiry may be demanded, the results of which will be placed before the Scottish Parliament.

A National Park Authority (NPA) is established to ensure that the National Park aims are collectively achieved in a coordinated manner. It comprises five locally elected members, ten ministerial, and ten local authority appointments.

To date, two National Parks have been designated in Scotland: the Loch Lomond and Trossachs, and the Cairngorm. These parks cover in total 5665 km^2 of some of Scotland's most dramatic landscape and varied habitats: lochs and rivers, agricultural lowlands, and uplands, including the largest arctic montane environment in the UK. Both parks attract a wide range of visitors including walkers, climbers, sailors, winter sports, and field sports enthusiasts, to the continuing economic benefit of these areas.

Land ownership within the parks is by individuals and charitable and commercial corporate bodies. No land is owned by the NPAs, which essentially have a legal planning function. The NPA is the planning authority in Loch Lomond and Trossachs, but in Cairngorm it shares the function with four local authorities.

The challenges presented by the National Park function

Balancing economic considerations with conservation and other aims, and the expectations of different stakeholders, requires a sensitive approach. Some forms of recreation conflict not only with existing land use but also with other recreational activities. The following challenges apply to both parks in varying degrees depending on the nature of the environment and the type of land use in operation or proposed.

Landscape strategy issues

The NPAs are required to use the planning controls at their disposal to ensure that landscape values are maintained without curbing economic or recreational development. Landscape is often taken for granted until change is proposed, for example through extension/development of the built environment, mineral extraction, wind farms, and power transmission lines. Also of concern are new forestry plantations, hill tracks to support deer stalking, or footpaths used by walkers and mountaineers which become highly visible as a result of heavy use.

Biodiversity Action Plans

The park authorities are required to address biodiversity issues within the context of the Scottish Government Biodiversity Strategy. This requires sustained effort to maintain and improve existing diversity as well as considering the appropriateness of reintroducing species that have been lost. Such issues are not easy to resolve, for human impacts over the millennia may have completely negated the opportunity for reintroduction. Also, there is now the complicating factor of climate change (see Chapter 10). Research is needed to enable predictions to be made regarding the speed and impact of climate change on animal and especially plant species.

A large number of actual and potential issues need facing. These include: how will efforts to encourage an extension of the natural pine forest impact on the legitimate field sports of stalking and grouse shooting? (See Box 12.1 for the bird and other animal species mentioned here and in the rest of this section.) Can the regeneration of Scots Pine be achieved without reducing the numbers of deer and sheep to a level that compromises their financial viability? Scotland has recently lost nearly 50 per cent of its heather cover and to keep what remains healthy and productive for deer, sheep, and red grouse calls for careful management including, where appropriate, a patchwork of controlled burning (known as the Muir burn). Can such activity help to protect areas of regenerating woodland? Will the bold experiment being carried out within the Cairngorms National Park succeed in achieving regeneration of native forest through the reduction of deer numbers (a cull has reduced red deer from 5000+ to 2000+) without having to exclude deer with fences? Does the considerable reduction of deer numbers in one area encourage deer to move in from neighbouring areas? And what is the impact on the financial viability of these estates? Will more deer fences mean more bird deaths through colliding with them, particularly by the increasingly endangered capercaillie?

BOX 12.1 BIRDS, SQUIRRELS, AND BEAVERS IN SCOTLAND'S NATIONAL PARKS

Four species of bird mentioned in the text are:

- Ptarmigan, a high mountain species currently hunted but possibly at risk from climate change;
- Red grouse, a heather moorland species managed for hunting;
- Black grouse, which occurs in areas containing moorland and woodland and is protected;
- Capercaillie, which relies on coniferous forest and is a protected species. The regeneration of forest with native Scots Pine is important to this species.

The grey squirrel was introduced to Scotland from North America. It competes with the native red squirrel which it has banished from many areas. A programme to reduce the numbers of greys is currently being considered.

The European beaver has been extinct in Scotland for 400 years. Beavers have recently been reintroduced into Scotland under a pilot scheme.

Each different habitat has its challenges. In the woodlands, how might the issue of the grey squirrel's fatal impact on the native red squirrel be resolved? How are the reducing numbers of black grouse and capercaillie to be addressed? What impact will climate change have on high upland species of ptarmigan and an array of alpine plants? What will be the impact of the reintroduction of the formerly native beavers?

All these unanswered questions illustrate that understanding of biodiversity remains limited and that research is an essential part of delivering the NPAs' Local Biodiversity Action Plans.

Access plans

The Land Reform Act 2000 sets out the rights and responsibilities of those accessing the land and lochs on foot, horse, or non-motorized boats for recreational purposes. Access can impact negatively on efforts to maintain or encourage biodiversity if not carefully considered. An Access Code allows those responsible for conservation activities to prevent or control access when species are put at risk; this particularly applies seasonally to breeding animals. There can also be conflict between those taking access and economic uses of land and water. Arrangements to reduce such conflict are needed.

Also required under the Act is the development by local authorities of a network of footpaths round communities. It is important that the NPA is involved in their planning and bringing biodiversity concerns to the forefront.

Education policy

Education and information are essential for both residents and visitors. The enjoyment gained from a visit should not be at the expense of those who manage the land for economic gain, conservation, or other reasons. Park Rangers are employed to welcome visitors and promote access, but they are fragmented across the various private estates within the Park. The NPA should help extend and coordinate this resource and NPA information distribution networks should include local schools.

The above represent just a few of the challenges facing the NPAs as they seek to achieve their aims and ensure the future of the parks. Achieving balance between conservation and commercial activities is not easy. Understanding based on sound scientific information and sensitive management will encourage both residents and visitors to give the support and understanding that their national parks deserve to receive.

Uganda: conserving the tropical rainforests through community management

Reversing the current destruction of the world's tropical rainforests is a major challenge and the second case study in this chapter focuses on Mabira Forest reserve in Uganda. As a developing country, Uganda has been dominated in recent decades by multiple transitions, from war to peace, one-party rule to multi-party elections, and a government-protected economy to free markets and more liberal trade policies.

Uganda's development is conceived mainly in terms of economic development which results in enormous pressure on the land (especially tropical forests) and water bodies (especially Lake Victoria and the River Nile whose source is Lake Victoria). The forest cover is estimated to have shrunk from 10.8 million ha (35 per cent of Uganda's land area) in 1890 to less than 5 million ha (16 per cent of the land area) this century (Kayanja and Byarugaba 2001).

Currently, 70 per cent of Uganda's forest cover is on private land and 30 per cent on government land. Population growth (estimated at 2.5 per cent per year) increases the pressure on land, food, and energy resources. Many public institutions, for example schools, rely almost exclusively on firewood for cooking, as does over 90 per cent of the population. Areas are already experiencing shortages, and hence rising costs which have increased burdens on women and children who collect it for cooking. Over-harvesting by both logging companies and community users of the forest, pressure resulting from commercialization of other crops, poor planning and weak regulation by the government (Kayanja and Byarugaba 2001), and inappropriate technology by wood processing companies have resulted in the rate of exploitation becoming unsustainable (see also Chapter 3 this volume). There is limited institutional capacity and resources in both central and local government to improve planning and regulation, and little incentive for the private sector to comply.

Urbanization and industrial growth are also putting pressure on the forests. Many reserves on urban fringes are under threat, with almost 10,000 hectares lost in recent years. The most affected forest reserves are those close to the urban and industrial centres, including the subject of this case study—the 300 km^2 Mabira forest which lies alongside the main highway to Kenya, 54 km east of the capital Kampala and 20 km north of Lake Victoria.

Forest policy and management in Uganda: a brief history and the 'Save Mabira' campaign

A British Protectorate from 1892, Uganda gained its independence in 1962. In pre-colonial times, natural resources were stewarded by local indigenous people who were dependent on them for their livelihoods. From colonial times until recently, however, government policy has been based on the notion that local forest users are ignorant and destructive. In Uganda and elsewhere, state authorities responsible for policy making have looked down on the knowledge and capacities of local communities. In so doing they have overlooked the obvious, which is that these communities are the most interested parties in the sustainable management of forests that are their source of life.

Uganda's first forestry policy dates from 1929. Subsequent policies have alternated generally between stricter conservation and more liberal economic exploitation. The government's latest forestry policy, which was launched in 2002, witnessed a change, however, with respect to management. Previous policy reviews had assumed management by the state and were silent on the roles of other stakeholders, such as the private sector and rural communities, as well as on forestry's linkages with other sectors and land uses. The 2002 Forestry Policy acknowledged that the forestry sector encompasses a wide range of stakeholders whose interests are not fully addressed and whose roles and responsibilities need to be defined and coordinated. The policy was also informed by recognition that financial and human resources available to government forest departments were inadequate to carry out the task of policing. It therefore encouraged the active participation of local communities in the management of the country's forests.

The 2002 policy has also opened a, possibly unintended, political space for dissent. Taking its cue from action among forest peoples worldwide who have demanded respect for their rights through local, national and international alliances, a campaign was established to fight a major threat to the Mabira Forest reserve.

Legally protected since 1932, the reserve is home to many endangered species. It is one of the most biodiverse forests remaining in Africa and is regarded by Ugandans as part of our heritage and our children's future. The forest has been open to the public since 1996 for bird watching, forest walks, and other activities. It has cultural and historical value and is Lake Victoria's main catchment zone where it acts as a natural water filtration system. In recent years the forest has been managed by a local land committee comprising local government representatives and members who have been elected directly by the public.

The trigger for the campaign was the application to the Government in 2007 of the Sugar Corporation of Uganda Limited (Scoul), a subsidiary of the Indian-owned Mehta Group, to transfer about one-third of the Mabira Forest for sugar cane plantations. The President of Uganda, H. E. Yoweri Kaguta Museveni, and his cabinet supported the plan.

Both local and international environmentalists opposed it, arguing that it would destroy the rich biodiversity in the forest which is home to more than 300 species of birds, 200 types of trees and 9 different primates. They also feared increased soil erosion, damage to local livelihoods, and negative impacts on the water balance and regional climate. Despite these protests, the government agreed a 7,100 ha sugar cane farm be established on the site, arguing that the investment would generate 3,500 jobs and contribute 11.5 billion Ugandan shillings to the treasury.

Concerned about the effect on their forest-based livelihoods, about 1,000 members of the local community joined environmental activists in a demonstration against the plantation proposal in Kampala in April 2007. It ended in three deaths and there were riots against Asians, Scoul plantations were set on fire, and calls to boycott Scoul's Lugazi sugar circulated.

The President defended the deforestation plans. The local press quoted him as saying that he would 'not be deterred by people who don't see where the future of Africa lies' and that the activists did not 'understand that the future of all countries lies in processing' (http://en.wikipedia.org/wiki/Mabira_Forest).

Domestically, however, the planned land give-away had proved unpopular, though often less for environmental reasons than economic, political, and racial ones. Despite the President's support for the plantation, the deforestation plans were suspended to calm the increasingly unfriendly public response.

Various institutions joined the Save Mabira campaign and offered alternative land for sugar cane production. These included the Baganda cultural king and the local Anglican Church. (Baganda is the largest of the traditional kingdoms in Uganda, and the largest ethnic group in the country.) Since then, the government has gradually changed its perspective and recognized the rights of local people in managing the forest. The communities have also discovered a previously latent wish to take charge of their resources and their ability to raise voices when their interests are threatened.

Thus, although it took a campaign which turned violent to realize it, community-based forest management is being actively promoted in Uganda as one of the options for promoting sustainable use. Two main challenges need addressing, however:

1. At present there are only isolated examples of community-based forest management in the country as a whole. How can they survive within a context where powerful actors—transnational corporations, national governments, and international environmental organizations—have interests?
2. Building internal cohesion and strength within the forest communities. For example, how can the participation of women, who have specific needs, perspectives, and roles, be encouraged and maintained? Their active participation in decision making and the equitable sharing of benefits between men and women are crucial for ensuring the long-term sustainability of community-based forest management. Equally important is the need to generate the necessary conditions to promote the active participation of local youth, representing the future of the community.

With the above challenges addressed, community-managed forests have significant potential as an approach to balancing conservation and claims for economic development.

Conclusion

Scotland and Uganda are two relatively small countries, the former administered as an integral part of the United Kingdom but with devolved powers in many aspects of its governance. The latter is a self governing republic within the Commonwealth of Nations. Both have international obligations to a variety of treaties and not least to those in respect of biodiversity.

Although the specific issues differ, both countries have to balance conservation considerations with commercial interests, particularly when such interests conflict directly with the task of sustaining or developing biodiversity. The main general difference between the two countries, however, is that in Scotland there are well-established institutional processes in place for participation of different stakeholders and for resolving disputes, while in Uganda such processes are implicit in the 2002 Forestry Policy, but their implementation is underdeveloped. Thus, disputes can become violent with tragic consequences.

That said, in Uganda there is now a far better understanding of the contribution that local people can make to local biodiversity issues. It is important that governments of all nations provide systems that allow local voices to be heard with respect to development proposals, both large and small, that impact on their lives and their ability to contribute in a sustainable way to the nation's conservation obligations.

▬ SUMMARY

- Conservation issues affect both developed and developing countries. This chapter has examined these issues in the context of National Parks in Scotland and tropical rainforest in Uganda.

- Conservation has three main dimensions, relating to social/cultural heritage, the physical environment, and local livelihoods.

- In Scotland, trade-offs are necessary between conservation of a dramatic landscape with accompanying biodiversity of both plants and animals, and use for economic activity, including public access and associated activities.

- In Uganda, a major issue is the tension between conservation of tropical rainforest with its abundant biodiversity and support for traditional livelihoods, and its destruction to allow commercial, plantation agriculture.

- Careful management involving multiple stakeholders and interests is required in both cases. In Scotland, the National Parks provide a framework in which the necessary trade-offs can be considered and negotiated. Recent policy in Uganda has also recognized the importance of more inclusive decision making and that forest communities comprise an important stakeholder. Enactment of inclusive processes remains, however, an issue, and the case study highlights an unruly process with respect to the Mabira forest which became violent.

■ ACKNOWLEDGEMENTS

The section on National Parks in Scotland acknowledges the following sources:

Scottish Natural Heritage (2006) Access Code
Biodiversity http://www.biodiversity.scotland.gov.uk
Cairngorm National Park http://www.cairngorms.co.uk
The Land Reform Act http://www.opsi.gov.uk/legislation/scotland/acts 2003
Loch Lomond and the Trossachs National Park: http://www.lochlomond-trossachs.org
National Parks (Scotland) Act http://www.opsi.gov.uk/legislation/scotland/acts 2000

The section on Mabira forest in Uganda acknowledges the following sources:

The World Rainforest Movement (2004) *Community Forests: Equity, Use and Conservation.* Available from http://www.wrm.org.uy/subjects/CBFM/text.pdf
Sustainability Watch, the Southern Civil Society Network. Available from http://www.suswatch.org/uganda/

■ REFERENCES

Kayanja, F. and Byarugaba, D. (2001) Disappearing forests of Uganda: the way forward. *Current Science* 81(8): 936–947.

13

Managing water for sustainable development across international boundaries: the Rhine river basin

Sandrine Simon

Introduction

Water resources are one of the most critical environmental issues of the twenty-first century. Water is an essential ingredient for life, for direct consumption, for use in industry and agriculture, for ecosystems and climate systems, and for the amenity and spiritual services it offers. Demand for water increases as the world's population rises and further demand is made by people's activities aimed at maintaining or improving current living standards. The stock of fresh water accessible to humans on Earth is limited, and meeting these growing needs becomes increasingly difficult. This shortage, of one of the most vital natural resources, is creating tensions and competition that have the potential to escalate into disputes and even armed conflicts.

The world water crisis, and conflict related to it, has attracted most attention in water-scarce areas, for example the Middle East and parts of India and Africa. Europe, in contrast, has the second-highest quantity of internal fresh water resources per capita (11,122 m^3) after Latin America (24,702 m^3). This can be compared with the Middle East/North Africa with only 761 m^3 per capita (data for 2004) (World Bank 2007: 106–107). One could conclude that Europe is, comparatively, exempt from water security issues.

However, in Europe, the issue is not so much the quantity but the quality of the water. Polluted water is a very important contributor to the world water crisis. It considerably diminishes the availability of safe, usable water and can severely damage water ecosystems that are essential for the quality of water habitats and of the water itself. The intense industrialization processes typical of economic development in most European countries have been a major contributor to the pollution of water systems over the years. In the case of shared rivers, where pollution by upstream countries can negatively affect downstream countries, this can lead to disputes between water users.

The river Rhine (Figure 13.1) on which this chapter focuses, illustrates this fact particularly well. Shared by nine countries (the Netherlands, Belgium, Germany, France, Luxembourg, Switzerland, Italy, Austria, and Liechtenstein), the Rhine river basin is the most industrialized region and the busiest waterway in the world. Vast industrial complexes are found along the river, such as those in the Ruhr, Main, and Rijnmond areas, which include Europe's biggest chemical production plants. Its waters are also used in agricultural production, energy generation, the disposal of municipal wastewater, recreational activities, and as drinking water for more than 20 million people. Furthermore, it constitutes a natural habitat for many plants and animals.

The water of the river is used for many purposes by the countries that share it. For instance, Germany, in which most of the river is located (56.5 per cent), uses the river for sewage disposal, navigation, industrial processes, drinking water, and amenity purposes. The Netherlands, with 6 per cent of the river, also has uses for hydropower, irrigation, and fishing. In ecological terms, these multiple uses are expressed through the concept of 'environmental functions' (see Table 13.1). Water resources perform these functions by providing goods (e.g. quantity of water, animals and plants living in water habitats) and services (e.g. waste recycling provided by ecological processes) that ultimately help to meet human needs.

Even though, in principle, a river can carry out all environmental functions simultaneously, this depends on ecological thresholds (for dilution, or renewal) not being exceeded. Activities that negatively affect the ecological functions of the river may be unsustainable. For instance, the sink function (Table 13.1) can be performed only if there is sufficient water to adequately dilute incoming pollutants. There is a point beyond which natural abatement cannot happen and at which all environmental functions are affected: water is not safe for drinking, habitats are damaged, fish cannot survive. This is what happened to the river Rhine, whose near-death is described in the next section.

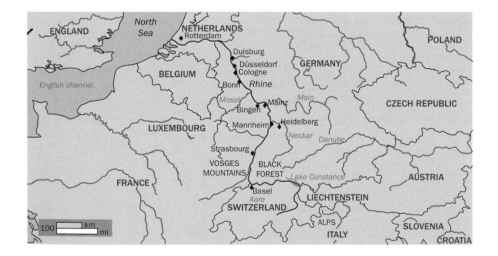

Figure 13.1 The Rhine basin.

Table 13.1 Environmental functions of water.

Source functions	Sink functions	Life support functions	Human health and welfare functions
Water catchment and groundwater recharge.	Regulation of the chemical composition of oceans.	Fulfilment of water habitat requirements.	Purification of water then usable for human consumption.
Water for drinking, irrigation, industry.	Dispersal and dilution of pollutants.	Regulation of run-off and flood protection.	Water for recreation.
Medium for transport.			Aesthetic, spiritual, religious, educational, cultural source of inspiration.

Source: Ekins and Simon (2003).

Managing a river in a sustainable way implies carrying out water-dependent activities while protecting these multiple functions. This can be enabled through the creation of appropriate **governance** systems that address conflicts head on and which value conflict management as a learning experience and an aid to decision making processes. This happened in the Rhine river basin in which, as described in the later part of this chapter, riparian countries (i.e. countries through which the river flows; riparian = of the river bank) negotiated the rebirth of their river through the creation of umbrella institutions.

The problems and disputes caused by unsustainable practices

The slow death of the Rhine started with the rapid industrialization and population growth of the mid-1800s. The various 'illnesses' of the river emerged from the developmental options chosen since that time by the countries that share it. The pollutants contributing to the destruction of the river included agricultural run-off (now the major source of pollution in the Rhine) and a long list of others such as pesticides from Swiss chemical factories, potassium salts from Alsatian mines, heavy metals from German industries, raw sewage from Rhineside cities, and engine oils from cargo ships.

Pollution affected the use of water for drinking purposes as well as fish stocks and the river habitats used by migratory birds and other species. The natural habitats had already been damaged throughout the nineteenth century by a series of physical alterations to the river such as construction of canals and diversions of the river for transportation and industrial purposes. Over time, the Rhine became 'Europe's cesspit'. The fish population dwindled and by the mid-twentieth century, salmon had vanished entirely (Weber 2000). Some examples of the problems affecting the Rhine are given in Table 13.2.

The problems outlined in Table 13.2 have inevitably given rise to disputes among and between different users of its waters. The most noticeable dispute in the Rhine basin was perhaps the series of protests carried out by the fishing industry when the salmon disappeared from what was once the most important salmon river in Europe (Swanson 2001).

Table 13.2 Water-related problems in the Rhine basin.

Main focus	Description
Navigation	Historically, even from Roman times, the Rhine formed the most disputed border between people in north-west Europe. Physical obstacles hindered navigation and tolls had to be paid.
Salt	The delta of the Rhine is affected by salt intrusion from the North Sea but also from large industrial plants in the hinterland. The salt in the Rhine's water affected its uses for drinking as well as for agricultural purposes in the Netherlands.
Industrialization and population growth	After 1850, the rapid industrialization and population growth generated problematic organic and inorganic pollution in the river. Agricultural run-off is now the major source of pollution in the Rhine. Downstream countries are affected by upstream polluting activities and countries do not necessarily agree on water quality criteria.
Fisheries	Until 1900, fishing was an important activity along the Rhine, so much so that stocks of salmon had to be protected against over-fishing and measures had to be undertaken to ensure an equal distribution of the catches along the river. A pronounced emphasis on navigation and hydropower shifted the focus of activity and progressively altered the fish habitat and, in turn, the fisheries industry.
Accidents	In 1986 a warehouse at the Sandoz chemical factory, near Basel, Switzerland caught fire: 1000 tonnes of agro-chemical substances were destroyed and 15 million litres of water were used to extinguish the fire, much of which mixed with the chemicals and flowed into the river. This caused the death of all aquatic life downstream.
Ecology	The physical alterations of the river changed the natural habitats of the Rhine throughout time. In addition, various pollution accidents affected these habitats with damaging consequences for fish, birds, other animals, and plants.
Flooding	Flood damage has been caused by physical modifications, including interventions to prevent floods which have done little more than displace the problem to another section of the river. Overall the risks of flooding have increased. Recent great floods of the Rhine occurred in 1993 and 1995 when dikes in the Netherlands were at risk of bursting.

Adapted from UNEP (2002), Frijters and Leentvaar (2003) and International Commission for the Protection of the Rhine (2008).

The return of the salmon, a target which was actually achieved by 2000, required many interrelated conditions to be fulfilled. The three-phase, 15-year 1987 Rhine Action Plan for Ecological Rehabilitation (known as the Salmon 2000 project) was aimed at reviving the river by identifying and eliminating the main sources of pollution and included water-quality testing, pollution patrols, and penalties for polluters. The rationale was to manage the restoration of the river by working on the premise that it is a *total* ecological system: water quality standards have to be respected, habitats restored, alterations to the physical flow and banks for flood control and navigation purposes have to be engineered in more ecologically friendly ways—and all at the same time. Whilst the disappearance of the salmon implied a total collapse of integrated ecological services, its return was therefore the symbol of the recovery of a whole river system.

The plan illustrated a growing awareness of the need to put into practice the concept of integrated water resources management (IWRM), defined by the Global Water Partnership as 'a process which promotes the coordinated development and management of water, land and related resources to maximize the resultant economic and social welfare in an equitable manner without compromising the sustainability of vital ecosystems' (Cosgrove 2003: 81). IWRM advocates taking account of the interrelations between water uses and water ecological services, as well as creating better coordinated decision making and action amongst stakeholders who share water resources. In the case of the Rhine river basin, the former translated into integrated water management plans, agreements, treaties, and conventions, whilst the latter led to the creation of umbrella river basin institutions, after years of fragmented actions by the various countries involved, as the next section shows.

> The Global Water Partnership's (GWP's) vision is for a water secure world. Its mission is to support the sustainable development and management of water resources at all levels.
> GWP was founded in 1996 by the World Bank, the United Nations Development Programme (UNDP), and the Swedish International Development Agency (SIDA).
>
> Global Water Partnership (2009)

Institutional, legal, and scientific responses for conflict management

Responses to the conflicts over the Rhine water resources were initiated centuries ago, often through umbrella river basin organizations representing some or all of the riparian countries. For example, in 1815, a Central Commission for Navigation on the Rhine (CCNR), comprising the Netherlands, Belgium, Germany, France, and Switzerland, signed up to maintenance of a uniform legal regime governing navigation. The first salmon treaty was in 1869, the most recent in 1987 (for the protection of the salmon and other species). Other treaties have concerned flood control, pollution control, and river diversions. One landmark was in 1950, when the International Commission for the Protection of the Rhine against Pollution (ICPR) was created. Among other initiatives, the Rhine Alarm Model was created in 1990 with alarm stations established along the river for rapid response to pollution incidents. More recently, in 1998, the 12th conference of Rhine ministers adopted guidelines for a programme on sustainable development taking into account improved protection as well as the uses of the river. This century has seen more holistic approaches agreed, such as the programme Rhine 2020 which focuses on improving the river ecosystem, flood prevention, groundwater protection, and continued monitoring.

Many of the earlier agreements focused on one particular problem (e.g. navigation, or pollution from one particular type of pollutant) and were also characterized by reactivity rather than prevention. It was only after the 1986 Sandoz chemical factory accident (see Table 13.2) that more preventative and integrated approaches to water resource management were developed and that real progress was made to speed up the cleaning of the Rhine. It thus effectively took an ecological catastrophe for the riparian countries to realize the importance of a long-term, preventative, integrated management approach.

However, while creation of overarching umbrella institutions has proved important for successful, coordinated water management along the Rhine, it represents only half the battle and can bring with it new conflicts, such as conflicts over which water quality criteria to use or technical monitoring method to adopt. More fundamentally, the original problems of widespread pollution, fisheries in trouble, and destroyed habitats remain if the umbrella institutions only address particular problems. The post-1986 period illustrates that it was only once the countries' coordinated efforts and knowledge were complemented by an integration of the issues at stake that the problems were genuinely tackled and the concept of conflict management and prevention started to make real sense.

This integrated approach is the aim of the amended ICPR, first established in 1950, which now addresses all of the interrelated ecological, economic, and social dimensions of what constitutes 'sustainable management' of the river. Also, there has been a ministerial declaration on a coordinated implementation of the European Union (EU) Water Framework Directive for the entire Rhine catchment in 2000, encouraging the operationalization of the concept of IWRM. Enriching even further the concept of integration, a 1999 Convention on the Protection of the Rhine stipulated that information must be exchanged with environmental NGOs and local communities whose knowledge, interests, and needs also need to be taken into account.

Conclusion

Have the Rhine countries found a strategy for managing their water resources in a sustainable way? If doing so translates into 'preserving the ecological integrity of water supply systems, wasting less water, allowing fair access to water supplies and giving people a say in how water resources are developed and used', as Miller (1998:511) advocates it should, then we can conclude that the efforts of the Rhine riparian countries are going in the right direction. In their case, however, the sustainability characteristic of their strategy must also take account of the fact that the river is shared.

As history has shown, reactive measures undertaken to clean up damage caused by pollution – often taken by reluctant heads of states – succeeded each other without altering the industrial activities in the area. The Sandoz ecological catastrophe and the downstream fisheries protests against the death of the river triggered the creation of more coordinated and integrated approaches to the management of all resources at stake, complemented by sophisticated technical monitoring systems.

The near death of the Rhine illustrated the risks generated by the type of economic activities carried out in industrial Western Europe. It showed that not only do ecosystems have limits in terms of their capacity to absorb damage generated by economic activities, but also that economic strategies can be negatively affected by the damage caused to the ecosystems, and that disputes over the resources can arise. Ultimately, economic activities depend upon the natural environment and the more intensively they use it, the more actively it should be protected. The so-called Rhine alarm systems might reveal that the levels of industrialization in the region are simply not viable.

These insights might help with other trans-boundary environmental issues – as ultimately most of them are. There is potentially one major shortcoming. The riparian Rhine countries had the financial, technical, and institutional means to develop new river basin organizations and new integrated approaches. However, many countries would simply not be able to afford the costs of repairing the damage caused by such an intensive type of economic development which, whilst often described as economically successful, can be ecologically disastrous.

■ SUMMARY

- In order to prevent conflicts from arising amongst stakeholders with competing water needs, many factors need to be considered together including water pollution, biodiversity protection, habitat restoration, industrial and flood risks, and agricultural impacts.

- The Rhine-sharing countries progressively developed an integrated management approach using an institutional framework. This was consolidated by technical monitoring support facilities and aimed at promoting political negotiations and the creation of legal agreements.

- While this institutional approach is often described as a success, the adoption of economic practices based on intense industrialization that are at the root of the original near-death of the river, still needs to be addressed critically.

■ REFERENCES

Cosgrove, W. J. (2003) *Water Security and Peace: A Synthesis of Studies Prepared Under the PCCP – Water for Peace Process. UNESCO PC-CP Series*, no. 29. UNESCO: Paris.

Ekins, P. and Simon, S. (2003) An illustrative application of the CRITINC framework to the UK. *Ecological Economics* 44(2–3): 255–275.

Frijters, I. D. and Leentvaar, J. (2003) *Rhine Case Study. UNESCO, IHP-VI. Technical documents in hydrology. PC-CP series*, no. 17. Paris: UNESCO.

Global Water Partnership (2009) http://www.gwpforum.org/accessed February 2009.

International Commission for the Protection of the Rhine (2008) http://www.iksr.org, accessed October 2008.

Miller, G. T. (1998) *Living in the Environment. Principles, Connections, and Solutions*, 10th edn Belmont, CA: Wadsworth.

Swanson, P. (2001) *Water, The Drop of Life*. Minnesota, MN: Wadsworth Press.

UNEP (2002) *Atlas of International Freshwater Agreements*. Nairobi: Nairobi.

Weber, U. (2000) The 'miracle' of the Rhine. *The Courier* UNESCO. Available from http://www.unesco.org/courier/2000_06/uk/planet.ht

World Bank (2007) *Collins Atlas of Global issues. A Visual Guide to the World's Greatest Challenges*. Washington, DC: International Bank for Reconstruction and Development/World Bank.

■ FURTHER READING

Burchi, S. and Spreij, M. (2003) *Institutions for International Freshwater Management. UNESCO, IHP-VI. Technical Documents in Hydrology. PC-CP Series*, no. 3. Paris: UNESCO.

Newson, M. (1997) *Land, Water and Development. Sustainable Management of River Basin Systems*, 2nd edn. London: Routledge.

Ohlsson, L. (1995) *Hydropolitics, Conflicts over Water as a Development Constraint*. London: Zed Publications.

Wenger, E. (2004) Integrated water resources management—experiences from the Loire and the Rhine. Fifth European Regional Meeting on the Implementation and Effectiveness of the Ramsar Convention. Available from http://www.ramsar.org/mtg/mtg_reg_europe2004_docs1b4.pdf.

■ USEFUL WEBSITES

Website of the International Commission for the Protection of the Rhine. **http://www.iksr.org/index.php?id=295** A comprehensive site giving information on the ecology, geography, and economic aspects of the basin as well as details on the structure, functioning, objectives, and strategies of the ICPR.

BBC Education: The Rhine. **http://www.bbc.co.uk/scotland/education/int/geog/eei/rivers/rhine/casestudy.shtml** This site explores, through maps, animations and exercises, river pollution and the various strategies adopted on the Rhine.

UNESCO's *From potential conflict to co-operation potential* (PC-CP) case study on the Rhine: **http://www.unesco.org/water/wwap/pccp/summary/cs_rhine.shtml** This site gives you the opportunity to contextualize and compare the type of conflicts taking place between Rhine riparian countries with those occurring in other river basins.

The United Nations Environment Programme site on Freshwater resources in Europe: **http://www.grid.unep.ch/product/publication/freshwater_europe.php.** This site provides information about various European watersheds and helps with contextualizing the Rhine case study.

14

Municipal waste management and the environmental health challenge: a tale of two cities

Stephen Burnley, Richard Kimbowa, and Isaac Banadda Nswa

Introduction

In this chapter we consider municipal waste management in two similar-sized cities; Kampala, the capital city of Uganda, and Birmingham, the United Kingdom's second-largest city. The amount of waste produced in the two cities and the way the waste is managed are compared. The differing priorities for the two situations are discussed in the context of sustainable development.

There are many definitions of municipal waste but we are defining it as *the solid wastes produced in domestic residences and similar wastes produced by commerce and industry*. Note that this excludes liquid wastes such as sewage, but in areas not served by a sewerage system, some human waste will be found in the municipal waste stream.

In our view, the principal objective of waste management is to protect the health of the public. However, in the highly developed Western economies it is easy to forget this. Waste management policies in these areas concentrate on resource recovery and the reduction of mid- to long-term pollution concerns including global climate change. For example, the European Union's aim is to make waste management more 'sustainable' by encouraging member states to manage their wastes in the following order of preference (Official Journal of the European Union 2008):

- Reduce—don't generate a waste in the first place
- Reuse—use products for their original design purpose more than once
- Recycle—reprocess waste materials to manufacture new products
- Recover—burn wastes to recover energy for heat and power generation
- Dispose—landfill wastes that are not suitable for recovery, recycling, or reuse.

These aims are very laudable, but make no mention of health protection and disease prevention—this hierarchy rightly assumes that EU members already have a system of waste collection and treatment that protects human health.

In contrast, waste legislation in developing countries places much more emphasis on the practicalities of collecting waste and of its health impacts. For example the Kampala City Council Waste Management Ordinance of 2000 concentrates on encouraging the safe storage of waste before collection by licensed firms. Similarly, in the western province of Sri Lanka, the Municipal Waste Management Rules focus on establishing collection schemes for waste, street cleanliness, and the establishment of sanitary landfills to replace open dumps (Waste Management Authority 2005).

Waste management in Kampala

Kampala occupies an area of 163 km^2 and had a resident population of 1,189,142 growing at a rate of 3.7 per cent per year in 2002 (the latest data available) (UBOS 2002). In common with many African cities, Kampala suffers from widespread poverty with 27 per cent of the people living in slums and other unplanned settlements: 38 per cent of the population are under 15 years old and only 1.8 per cent are aged 60 or above (UBOS 2002). In contrast, 23 per cent of Birmingham's residents are under 16 and 19 per cent are over 60 (Birmingham City Council 2007).

Kampala faces many urban management challenges including the provision of safe water, improved sanitation, and solid waste management. This is hindered by the rapid population growth, the proliferation of slums, and an inefficient local revenue raising system.

Waste collection and transport in Kampala are the responsibility of the five divisions of the city while Kampala City Council (KCC) is responsible for final disposal of the waste. In recent years, the increasing population, changes in consumption habits, and increased economic activity have all led to an increase in solid waste generation. As a result, some of the divisions have started to involve the private sector in waste collection activities. However, this involvement has focused on the affluent neighbourhoods, leaving the majority of the area unserved. This is mainly due to the ability and willingness of the affluent residents to pay for better amenities including a waste collection service.

Kampala's current population generates about 504,000 tonnes of waste per year of which only 35 per cent (180,000 tonnes) is collected (KCC 2006a). The collected waste is disposed of at the Kiteezi landfill, which is operated by a contractor under KCC supervision. The remaining waste is either fed to animals, spread on gardens, or burned by residents. Waste is also illegally dumped in the drains, swamps, and bushes. The composition of this waste is discussed below, but the main component is organic waste at 74 per cent of the total.

The Kiteezi landfill is about 13 km from the city centre on a 29-acre site owned by KCC (Figure 14.1). It is fitted with a clay liner and has a leachate treatment facility. The site opened in 1996 and was expected to last for 12 years. However, in 2008 KCC acquired a further 2.5 hectares adjacent to the current landfill, which is expected to extend the

Figure 14.1 Kiteezi landfill site.

existing life of the facility by at least another three years. As well as municipal waste, the site accepts animal carcases, sterilized hospital waste, sewage sludge, inert wastes (such as construction waste), and condemned foods (KCC 2006b).

> Leachate is the liquid that seeps out of the base of a landfill as the waste decomposes; it can contaminate ground and surface waters if not collected and treated.

The informal sector plays an important role in waste management in Kampala. Two main activities are carried out. Communities who live near the Kiteezi landfill scavenge the working site for reusable or recyclable materials such as cardboard, plastic, metals, and household items. There are also people in the city itself who collect reusable materials from the waste before it is collected. These people include plastic and glass bottle collectors who gather bottles from shops, restaurants and hotels, and farmers who pay market vendors for banana and potato peelings to be used as animal feeds. For Kampala's large organic waste fraction, a substantial amount is recovered in this way for the small-scale farmers with cows, goats, and pigs in and around Kampala, before the remainder is piled up for landfilling.

Future plans for waste management

As part of its review of the existing waste management strategy, KCC is considering appropriate waste processing technology options, which can reduce the tonnage of waste being disposed of in its landfill. This will extend the landfill life, but ignores the problem of low waste collecting rates. Despite this, the existing capacity of the landfill is inadequate, hence the need for extension of the site. In the long run, another solid waste disposal facility at a different site altogether is planned.

Waste management in Birmingham

The city of Birmingham occupies an area of 268 km² and in 2005 the population was 992,400 and growing at about 0.4 per cent per year: 60 per cent of the population lives in properties owned by the residents with the remainder renting property from the local authority, private landlords, or not-for-profit housing associations. All properties are connected to the water, sewage, and electricity systems.

Birmingham City Council is responsible for collecting household waste from all properties in its area and disposing of this waste in accordance with UK law. This service is paid for from council tax (a local tax based on the value of the property) and central government grants to the city. For the year 2006–07 the average household waste collection and disposal charges for Birmingham were £35.10 per tonne and £44.89 per tonne respectively (Audit Commission 2008).

Waste is collected from all of the city's households at weekly intervals. In addition there is a separate collection of paper, metals, glass, plastic, and garden waste for recycling or composting from around 89 per cent of the properties. In the year 2004/05 the city collected 501,000 tonnes of household waste plus 68,000 tonnes of waste from business and industrial premises (the remainder of the city's commercial and industry waste was managed by private firms under arrangements made directly with the businesses). This overall figure of 569,000 tonnes compares with the 180,000 tonnes collected in Kampala. The disposal routes for the household waste are shown in Figure 14.2.

Energy recovery is carried out in an incinerator that produces 25 MW of power (sufficient to supply around 37,000 UK homes). The incinerator was built in 1996 by the private company that currently operates the plant. Waste that cannot be recycled or incinerated is sent to a landfill site that incorporates control measures to collect any leachate and to burn the gas that forms as the waste decomposes.

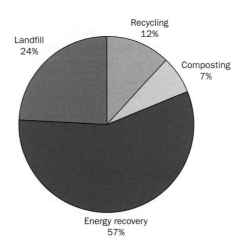

Figure 14.2 Management of Birmingham's municipal waste.
Source: Birmingham City Council 2006.

Both of these waste disposal facilities are regulated by the Environment Agency – the UK Government body responsible for ensuring that all waste collection and management operations meet national and European environmental standards.

Future plans for waste management

In November 2006, the city produced a waste strategy for the period 2006–26 (Birmingham City Council 2006). Under this plan, the city intends to minimize its reliance on landfill by reducing the amount of waste generated and increasing its recycling rate to 40 per cent.

Overview of sustainability concerns for Kampala and Birmingham

Birmingham and Kampala are typical examples of how cities in developed and developing countries respectively manage their wastes.

By global standards, Birmingham is an affluent city within the UK, where nationally the annual UK per capita **gross domestic product** (GDP) is approximately US$45,000 (2008). It has a well-developed waste collection infrastructure, waste disposal is highly regulated and expensive technology is used to minimize pollution from waste disposal plants. Waste-related health infections are largely a thing of the past; progress is measured in terms of increasing recycling rates and sustainability is concerned with safeguarding the environment in the coming decades and centuries.

In contrast, there is widespread poverty in Kampala (Uganda national per capita GDP is approximately US$360 for 2008) and the city is still working towards city-wide collection of household waste. Whilst Kampala does have a sanitary landfill site, pollution control measures at the site are basic and there are many unofficial open dumps around the city. Cholera and other waste- and sewage-related diseases are still present. Sustainability is far more concerned with what is happening to the citizens today and tomorrow.

The most striking comparison between the two cities is in the amount of waste produced. Each Birmingham resident produces 1.4 times as much waste as a Kampala resident, but when the amount of waste collected is considered, the local authority in Birmingham handles more than three times as much waste as Kampala's authority.

The composition of the municipal waste in the two cities is also very different, as shown in Figure 14.3. Birmingham's waste is typical of an affluent city with high proportions of paper, food, and garden waste and smaller, but significant quantities of metal and plastics. In contrast, Kampala's waste is predominantly food waste, but the absence of a city-wide sewerage system means that human waste is also present. Glass and metals (which are related to the purchase of food and drink) account for a much lower proportion of Kampala's waste. Both cities produce similar proportions of plastic wastes; in Birmingham much of this is food packaging and drink containers whilst in Kampala, this is due to the

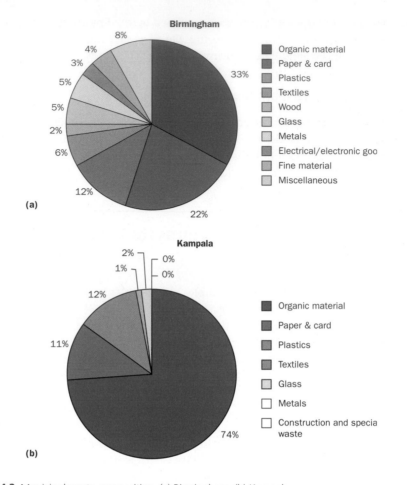

Figure 14.3 Municipal waste composition. (a) Birmingham. (b) Kampala.

Sources: (a) Birmingham City Council (2006); (b) Republic of Uganda (2006).

unrestricted and widespread use of different gauges of plastic carrier bags without any deliberate effort to recycle or reduce them, as well as the growing consumption of bottled water by those who can afford it.

So what should be the priority for waste management in these two cities? At the beginning of this chapter we suggested that protecting public health must be the first aim of waste management. If this is to be the case, the priority for Kampala should be to address from a health perspective the entire waste management chain from storage through to collection and disposal.

This view is widely accepted. For example, the process of developing Kampala's City Development Strategy (CDS), which was led by UN-HABITAT (United Nations Human Settlements Programme) identified sanitation, shelter, and income generation as key issues to be addressed (UN-HABITAT 2005). Waste management impinges strongly on all three areas. Amongst other effects, the lack of proper waste storage and management systems will:

- prevent local income generation by deterring the sale of foodstuffs and the establishment of other businesses;

- lead to blocked drains from uncollected waste undermining the quality of sanitation and hygiene;

- hinder individual, private, or public efforts to improve housing.

Sanitation, shelter, and income generation influence one another in similar ways. For example, better sanitation provides the chance for better shelter and vice versa, whilst the availability of better sanitation and shelter motivates development of related services and retail outlets (increasing income generation).

City Development Strategies (CDS) are action plans for equitable growth in cities, developed and sustained through participation, to improve the quality of life for all citizens. Kampala's CDS was developed between 2002 and 2005 and proposals based on action and investment plans to support the various intervention areas are expected.

Waste storage in Kampala is regulated by the Public Health Act (1964) and the KCC (Solid Waste Management) Ordinance (2000). These stipulate, respectively, that waste should be stored at its source in a dustbin, and prohibit its deposit in any place where it may cause a health hazard. Unfortunately, the 1964 Act is only weakly enforced and the 2000 Ordinance has not been implemented. Consequently, waste is dumped by the roadside, in drainage channels, and in other public areas. Residents of areas where there is no waste collection service have little option but to do this.

It would be highly desirable to provide a city-wide waste collection service with better enforcement of the storage regulations. However, this would involve higher costs with the more affluent residents having to subsidize the poor. Such a measure would almost certainly be highly unpopular with the wealthy residents. Furthermore, local revenue collection is highly inefficient (some sources suggest that only 20 per cent of the potential local revenue is received by the City Council) so raising local tax rates would be unlikely to provide the necessary funds for such a project. For both these reasons, better waste collection in the near future is regrettably an unrealistic hope.

What should happen to Kampala's waste after collection? Greater use of controlled sanitary landfills, such as the Kiteezi site, would add to the public health benefits and begin to address the environmental issues. However, even with the extension, Kampala is destined to run out of landfill capacity by about 2011. For this reason, KCC is planning to construct a new landfill site.

KCC recognizes the need to think about recovering resources from this waste as a means of extending landfill life and promoting a more sustainable waste management system. With its high organic content, biological processing is the most promising option. Anaerobic digestion is a relatively high-technology process but generates electrical power, whilst aerobic composting provides a lower cost option. Both processes destroy pathogens and produce a valuable product; the price of compost in Kampala is 50,000 Ugandan shillings (about US$30) per tonne (UEPF 2003).

Anaerobic digestion: microbial processing of organic waste in the absence of oxygen to produce 'biogas'— a methane/carbon dioxide mixture that can be burned to generate power and/or heat. The by-product digestate is a sludge that can be applied to soil to improve its structure, water-holding capacity, and nutrient levels.

Aerobic composting: microbial processing of organic waste in the presence of oxygen to produce a stabilized compost that can be used as a soil improver or blended with other components to create a growing medium.

The position of the informal sector also needs to be considered. Scavenging is an inherently hazardous occupation. As well as the danger to the scavengers, scavenging leads to condemned food from the landfill finding its way into poor households. Much scavenging activity is carried out by children and this contributes to absenteeism and drop-out from education due to the long hours spent scavenging. Simply banning all these informal activities would be unenforceable and would also result in some of the poorer residents losing their livelihoods. However, there are ways of bringing the informal sector into the mainstream to the benefit of both sides. For example, Fiehn (2007) advocates setting aside areas in landfill sites for scavengers to work, or using scavengers to collect recyclable materials from the kerbside. Wehenpohl and Kolb (2007) reviewed the work of the informal recycling sector in several countries and concluded that incorporating it into the municipal sector leads to improved conditions for the recyclers and higher collection rates at a lower cost for the municipality.

As for Birmingham, European and national legislation mean that the city has no choice but to work towards achieving higher recycling and recovery rates and greater reductions in the use of landfill. Economic instruments such as the £48 per tonne landfill tax (2010 rate), payable for every tonne of waste delivered to a landfill site, provide a further financial incentive for higher recovery rates. But should developed countries continue to go even further? In some situations a recycling rate of 70 per cent could be possible, but at a very high cost and arguably for only marginal environmental improvements. There is certainly a need for developed countries to quantify the environmental benefits and costs of their waste management strategies.

On a wider level, there is a clear need for all developed countries to set appropriate waste-related environmental standards. However, how far should they go? Principle 7 of the Rio Declaration, made following the United Nations Conference on the Environment and Development (**UNCED**) in Rio de Janeiro in 1992, calls on states to 'cooperate in a spirit of global partnership to conserve, protect and restore the health and integrity of the Earth's ecosystem' (United Nations 1992).

In the UK and other developed countries, waste management operations are among the most heavily regulated of all industrial processes, operating to stringent pollution control standards. The health and environmental benefits from adopting even higher standards would be marginal. Compliance with Principle 7 suggests that developed countries need to rethink their priorities, practices, and standards and consider putting a ceiling on waste management standards and recycling rates. Funding allocated to this area could then be mobilized and channelled through development agencies and NGOs to help raise standards in developing countries. Globally, this would lead to greater improvements in public health for a given expenditure and also have a positive impact on global climate change. Research should be commissioned to quantify the benefits of such a move and to evaluate the institutional and political challenges to its implementation.

■ SUMMARY

- In this chapter, we consider municipal waste management in two similar sized cities; Kampala, the capital city of Uganda and Birmingham, the United Kingdom's second-largest city.

- The quantity and composition of municipal waste in Birmingham and Kampala are typical of developed and developing countries respectively. Climate, affluence, lifestyles, and the provision of sanitation all influence waste generation.

- In Kampala, sustainability is measured in terms of the immediate effects on public health. To become more sustainable, Kampala needs to provide a city-wide waste collection service and secure further controlled landfill capacity.

- Birmingham is improving its sustainability by increasing waste recycling and recovery rates to address long-term issues and comply with national and European legislation.

- Research should be commissioned to evaluate the global benefits and barriers to the implementation of waste and other health/environmental protection measures in the developing world using funds diverted from environmental projects in developed countries

■ REFERENCES

Audit Commission (2008) *2006/07 BVPI data and quartiles*. Available from http://www.audit-commission.gov.uk/performance/dataprovision.asp, accessed 1 May 2008.

Birmingham City Council (2006) *Municipal Waste Management Strategy 2006–2026*. Birmingham: Birmingham City Council.

Birmingham City Council (2007) *Census 2001; about Birmingham*. Available from http://www.birmingham.gov.uk/GenerateContent?CONTENT_ITEM_ID=26205&CONTENT_ITEM_TYPE=0&MENU_ID=12672, accessed 18 July 2008.

Fiehn, H. (2007) *A case study on alternative approaches to waste characterization analysis amid salvaging and recycling issues in South Africa*. International Waste Management and Landfill Symposium, Cagliari, Italy.

KCC (Kampala City Council) (2006a) *Draft revised Solid Waste Management Strategy Report*. Kampala: Kampala City Council, Kagga in association with Kv3 (Kwezi V3 Engineers).

KCC (Kampala City Council) (2006b) *Kampala City Council Kampala Institutional and Infrastructure Development Project (KIIDP). Final Report prepared by Aquaconsult in association with EMA CONSULT LTD and SAVIMAXX LTD. Environmental Analysis*. Kampala: Kampala City Council.

Official Journal of the European Union (2008) *Directive on Waste and Repealing Certain Directives*. 2008/98/EC, L312: 3–30.

Republic of Uganda (2006) *Draft revised Solid Waste Management Strategy Report*. Kampala: Kampala City Council, Kagga in association with Kv3 (Kwezi V3 Engineers).

UBOS (2002) *2002 Uganda Population and Housing Census – Kampala District Report.* Available from http://www.ubos.org/onlinefiles/uploads/ubos/survey%20documentation/2002%20Census/survey0/data/Reports/Kampala.pdf, accessed 15 May 2008.

UEPF (2003) *Market survey of compost in Kampala City and its environs by Uganda Environment Protection Forum (UEPF).* A baseline survey for UEFP's Project: Integrated Approach to control Climate Change through Solid Waste Management (SWM) in Kawempe division (July 2002–June 2003), Unpublished.

United Nations General Assembly (1992) *Report of the United Nations Conference on Environment and Development: Annex I Rio Declaration on Environment and Development.* United Nations A/CONF.151/26 (Vol. I). Available from http://www.un.org/documents/ga/conf151/aconf15126–1annex1.htm, accessed 30 July 2008.

UN-HABITAT (2005) *Cities Development Strategies for Improvement and Poverty Reduction in the Lake Victoria region—Kampala, Kisumu and Musoma. A Documentation of the Process, Achievements and Lessons. Urban Management Programme—Working Paper Series.* Nairobi: UN HABITAT.

Waste Management Authority (2005) *Municipal Solid Waste Management Rules–2005.* Waste Management Authority (Western Province) Chief Ministry, Western Province, Sri Lanka.

Wehenpohl, G. and Kolb, M. (2007) *The economical impact of the informal sector in solid waste management in developing countries.* International Waste Management and Landfill Symposium, Cagliari, Italy.

■ USEFUL WEBSITES

WASTE advises on urban environment and development: **http://www.waste.nl/**. WASTE works towards sustainable improvement of the urban poor's living conditions and the urban environment in general focusing on bottom-up development in relation to recycling, solid waste management, ecological sanitation, and knowledge sharing.

Green Productivity—an approach to sustainable development: **http://www.apo-tokyo.org/gp/index.htm**. Green Productivity (GP) is a strategy for simultaneously enhancing productivity and environmental performance for overall socio-economic development. GP focuses on greening the supply chain, water resource management, energy efficiency, solid waste management, and integrated community development.

Urban Environmental Management: **http://www.gdrc.org/uem/waste/waste.html**. A rising quality of life and high rates of resource consumption patterns have had an unintended and negative impact on the urban environment—generation of wastes far beyond the handling capacities of urban governments and agencies.

Cities are now grappling with the problems of high volumes of waste, the costs involved, the disposal technologies and methodologies, and the impact of wastes on the local and global environment. These problems have also provided a window of opportunity for cities to find solutions—involving the community and the private sector; involving innovative technologies and disposal methods; and involving behaviour changes and awareness raising.

15

Environment, inequality, and the internal contradictions of globalization

Raphael Kaplinsky

Introduction

We are currently in the midst of a period of triumphalism about the unstoppability of deepening **globalization**, the belief that the forces of globalization are irresistible, that there will be no end to the ever-deepening and ever-widening removal of the cross-border barriers to the flows of products, production factors (people and capital, Box 15.1), ideas and technology. The view that globalization is unstoppable echoes that of an earlier era, the period ending with the First World War. The last decades of the nineteenth century represented a similar phase of rapidly deepening global integration. Yet it came to an end in an abrupt and brutal form, with the loss of many millions of lives, and it was only half a century later that we entered a new phase of global integration.

The primary reason for the descent of the global economy into war in the early years of the twentieth century was the failure of the old imperial powers to allow a new entrant—Germany—to play a key role in the forming of global architecture (Box 15.1). A similar challenge faces the global regime now that dynamic new entrants from the east are rising in the economic league and will soon be seeking to have this reflected in the role they play in the fashioning of the global political and institutional architecture.

This chapter addresses the sustainability of the global system, and will do so by focusing on three themes (Kaplinsky 2005). The first is to understand how the very nature of capitalist development leads to the degradation of the global commons (Box 15.1), to global warming, and to climate chaos. Second, it is also in the nature of current processes of globalization that inequality deepens and poverty endures, often reinforcing pressures on the environment. Third, this combination of environmental impacts and inequality sets up internal contradictions which are likely to undermine the sustainability of the globalization process itself. All of this leads us to question globalization–triumphalism and

---

BOX 15.1 ECONOMIC TERMS USED IN THIS CHAPTER IN ORDER OF FIRST APPEARANCE

- *Production factors*: the resources employed to produce goods and services.
- *Capital*: often used loosely to refer to a manufacturer's machinery and agricultural equipment, but more broadly is any form of wealth capable of being employed in the production of more wealth.
- *Global architecture*: the structure of the global economic and political system of firms, nation states, and their interactions which is governed by sanctions for those who either do or do not conform.
- *Global commons*: that which no one person or state may own or control and which is central to life. The two most obvious examples are the oceans and the air we breathe.
- *Command economies*: economies in which production (generally under the control of state-owned enterprises) was planned in detail centrally by the state. They are associated with communist party governments, in contrast to the capitalist market system.
- *Value chain*: the value chain describes the full range of activities which are required to bring a product or service from conception, through the different phases of production (involving a combination of physical transformation and the input of various producer services), delivery to final consumers, and final disposal after use.
- *Accumulation*: generally the amassing of objects of wealth; the increase in wealth or the creation of wealth.
- *Reserve army of labour*: those who are unemployed, but if given the chance, would seek to be employed; they are waiting to join the labour force,

Adapted from Wikipedia

---

requires us to think about new and innovative ways in which humankind can continue to survive and prosper, living with, rather than in command of, nature.

The innovation imperative, the environment, and global warming

There are many reasons why the command economies of the Soviet system collapsed, but perhaps the most important is that they failed to deliver the goods. That is, they neither grew as rapidly or delivered the quality and variety of goods of their capitalist counterparts. Joseph Schumpeter, an Austrian economist of the mid-twentieth century, provided the explanation for this systemic deficiency (Schumpeter 1942). Schumpeter showed how the very breathing of the capitalist economy requires **innovation** where entrepreneurs, who are defined by their capacity to turn ideas into profit, respond to competition by introducing new products and processes. Innovation and expansion are the core of the capitalist system, they are its internal motor.

Writing some centuries before Schumpeter, Adam Smith provided the key to understand how this accumulating motor of capitalism fuels a *globalizing* economy. Using an example of a pin factory, Smith showed how the division of labour led to an increase in productivity. Moreover, he argued, 'the division of labour depends on the extent of the market' (Smith 1776/1976)—that is, the bigger the market, the greater the division of labour, the greater the gains in productivity, and the higher the profit to the innovating capitalist.

Thus in summary, the argument for the inevitability of globalization runs as follows:

• Innovation and growth are at the heart of the capitalist system

• Capitalism triumphs because of its ability to innovate and grow faster.

• Increasingly, this growth takes a global form, as new large-scale technologies develop and new forms of firm and factory organization (of which the division of labour is an early example) result in increasingly global value chains (Box 15.1) producing for global markets.

This ever-expanding global system makes enormous demands on the environment because, amongst other things, global value chains and global markets require transport, and transport uses energy. There is no need here to go into the extent of these resource demands, nor the impact which this is having on the global climate. Forget for the moment the localized pollution which results from the ever-deepening exploitation of the earth's biosphere—the pesticides in cotton production, the cancer (mesothelioma) arising from asbestos production, the pollution of Alaskan waters through oil-spillages. As we are increasingly aware, these localized environmental impacts pale into insignificance when we see the rapidly growing impact on the global climate. We are not just in an era of global warming and climate change, but one of growing unpredictability and climate chaos (Sachs 2004).

The global capitalist accumulation (Box 15.1) system is making greater demands on the biosphere than it can sustain. For humankind to survive, to live in peace and in security, the accumulation motor needs to either be switched off, or perhaps even to be put in reverse. At the very least the energy-intensity of this innovation system needs to be reversed (Stern 2006).

The innovation imperative, and global poverty and inequality

The theory of comparative advantage which underlies modern theories about trade can be traced back to the writings of Adam Smith and David Ricardo in the late eighteenth and early nineteenth centuries (Smith 1776/1976; Ricardo 1817). By comparative advantage, Ricardo meant that countries should specialize in those activities in which they performed relatively well. The theory provides a particular perspective on global poverty, one which sees it as a *residual* phenomenon. That is, global poverty can be seen as a temporary condition, a condition which can be alleviated if all producers specialize in their areas of comparative advantage and enter the global system—the poor remain poor because they fail to join in. In the words of the World Bank, '[i]n sum, global economic integration has supported poverty reduction and should not be reversed' (World Bank 2002: xi).

There is, however, a key assumption in the intellectual architecture of this win–win approach to globalization of which David Ricardo was fully aware. Specialization in areas of comparative advantage only leads to a win–win outcome in a world of full employment, that is, if all producers have a role to play, a product to produce which someone else both wants and can afford to buy. What happens if this world of full employment does not exist?

Here we need to be informed by both Thomas Malthus and Karl Marx. Malthus, a British political economist writing in the same era as Adam Smith, argued that the growth of population would exceed the capacity of humankind to produce the necessities required to feed it. He was wrong of course in the sense that our *present* innovating and highly productive production system is clearly able to feed the world's *current* population (in principle, if not in practice). However, can it continue to do so on a *sustainable* basis, or will our demands for present consumption undermine our capacity to deliver adequate consumption in the future (Sachs 2004)?

Writing about fifty years later, Marx, too, had something to say of relevance to the win–win outcome to globalization. He argued that the technical progress which is key to the capitalist innovation system was inherently labour saving, and that this led to a systemic tendency towards a 'reserve army of labour' (Box 15.1) (Marx 1876/1970). In an environment of surplus labour, where productive capacities exceed consumption (in other words, where supply exceeds effective demand), there is a 'race to the bottom' for all those who do not have unique capabilities to offer. They are subject to the intensity of global competition and wages decline. In this race to the bottom, global poverty is not so much residual but *relational*, that is, a direct consequence of the workings of the global system (Bernstein 1990). It is not just that real wages and absolute standards of living may decline, but that the other side of the income spectrum has seen relative gainers. These are those able to insulate themselves from competition through special attributes that they are able to exploit to a global audience—high-end professionals, celebrities, sports people, innovators, and so on. Their incomes rise disproportionately in a global arena, leading to rising inequality, another form of poverty.

What do the facts show (for a detailed analysis see Kaplinsky (2005))? Let us begin with poverty understood as an *absolute* condition, focusing on the US$1 per day target which is the key **Millennium Development Goal**. The number of people globally living below the US$1/day level has remained stable since 1990 but rose between 1993 and 2006 in sub-Saharan Africa, Latin America, and South Asia (World Bank 2007).

Viewing poverty as a *relative* issue in relation to income distribution, the outcome has been unambiguous. In virtually every respect, as globalization has deepened, so the distribution of income within and between countries, regions, classes, and genders has become more unequal (Cornia and Court 2001; Kaplinsky 2005). China has also played an important role in the worsening distribution of income. With 20 per cent of the world's population, growing inequality in China has meant that, notwithstanding its rapid growth rate, the population-weighted global distribution of income has widened even further (Milanovic 2005).

Those people subject to global competitive pressures—unskilled workers and increasingly also semi-skilled and information technology (IT) workers—have experienced growing levels of unemployment and seen their relative incomes decline (Hira 2004). This has not just been a phenomenon experienced in low income developing countries, but in the rich countries as well, where income distribution has tended to worsen significantly over the past decade or two (Cornia and Court 2001). In the US (1966–2001), the top 1000th of the population gained more total growth than the bottom 20 per cent, and the top 1 per cent more than the bottom 50 per cent (Dew-Becker and Gordon 2005: 36). In the UK, the minimum wage is lower in real terms than it was two decades ago (Toynbee 2003).

Will things improve, will these global labour markets tighten? Almost certainly not. The total industrial labour force in the 14 largest OECD economies in 2002 was 79 million. In the same year, China's manufacturing sector employed 83 million (Kaplinsky 2005). More to the point, it is estimated that China's reserve army of labour—those waiting to enter paid employment and currently working in very low-productivity activities—was in excess of 100 million. To make matters worse, an increasing number of this labour force is educated and skilled, and if that was not bad enough, by 2020, India's labour force will exceed that of China. There are no signs of the global reserve army of labour drying up.

> The OECD (Organisation for Economic Cooperation and Development) consists of 30, mostly high-income countries which accept the principles of representative democracy and a free market economy.

What does this portend for the sustainability of globalization?

What do these observations about the environment, climate, and income distribution have to tell us about the sustainability of globalization? They suggest that globalization suffers from two internal contradictions, that is developments which arise from its very success but which at the same time threaten its very future.

First, the current trajectory of continued global growth is quite clearly unsustainable in environmental terms (Stern 2006). The energy required to transport products in extended global value chains is placing impossible demands on the planetary biosphere. If humankind gets round to taking action to stop global warming and climate chaos, this will necessarily have to be at the expense of the current trajectory of energy-intensive production systems, probably through much higher prices for energy. The logic of shipping low-value vegetables, fruit, and components around the world will be lost and profitable production systems will necessarily place a greater emphasis on proximity. Food miles, for example, are rapidly ascending in consumer consciousness as they ponder the rapidly evolving change in global climate patterns.

Second, globalization forces alterations in economic specialization. The result is frequent and significant change in employment patterns, in work organization, and in institutional design. Perhaps more importantly, it has also led to significant changes in the pattern of income distribution (see above).

The consequence is that life appears to have become more insecure for many, including for articulate professionals in the high-income economies. To a significant extent this growing anxiety and unease is a direct consequence of the imperative for continual reinvention forced by global competition. For example, during the late 1990s, the managers of General Electric subsidiaries (one of the largest US conglomerates) were expected to evaluate and weed-out the least-well-performing group of employees on an annual basis however competent they were in performing their allocated tasks. In the early years of the millennium, GE promoted a '70:70:70 policy'— 70 per cent of activities to be outsourced; 70 per cent of this outsourcing to be offshored (that is, sent abroad); and 70 per cent of this 'offshoring' to go to low-wage economies. It is an agenda of uncertainty, distrust, and fear. This is echoed in the worldview of the former head of Intel, Andy Groves, who wrote a best-seller entitled *Only the paranoid survive* (Groves 1996). In each case the prognosis was change—'reinvention', 'reorganization', 'business process engineering'—an ongoing agenda not just in the private sector but even in state-owned bureaucracies such as the UK's National Health Service and educational systems. This world of insecurity, fear, and anxiety threatens to engender opposition to globalization, the more so as the professional classes in the high-income economies are now being threatened by the offshoring of their own jobs to India and other lower-wage economies.

The lessons of the nineteenth century provide an important backdrop in understanding these possible developments in the early twenty-first century (Williamson 1998). After five to six decades of growing global integration, the world economy turned inwards after 1914 with the outbreak of the First World War, and the outward momentum was only regained in the decades after 1950. In between saw a period of inward focus and a reduction in economic integration. This reversal of global processes followed directly from the success of late-nineteenth century integration. Cheap grain imports into continental Europe led to a decline in agricultural profits. This resulted in the imposition of tariffs against agricultural imports in much of Europe. Second, there was a mass migration of unskilled Europeans into the United States (US) as 60 million people, often literally walking across Europe, made their way to the US between 1820 and 1914. This forced down wages in North America, and led to growing controls against migration. At the same time the competitiveness of US manufacturers threatened the survival of European manufacturing. This resulted in the imposition of tariffs against imported manufactures, and set in train a series of 'beggar my neighbour' policies in which countries reacted against each other by ratcheting up protection against imports. Finally, the demand for growing markets and resources led to the expansion of colonialism, which spurred the imperialist rivalries which helped to fuel the First World War. In each case, the seeds of change are to be found in the workings of the nineteenth-century global economy, and arose as a direct consequence of its success.

We can also see an interaction between the environmental, social, and political internal contradictions of the contemporary capitalist production system. Pushed to the margin of subsistence by the unequalizing and impoverishing nature of globalization, populations encroach further on fragile physical environments, exacerbating pressures on local ecologies. This leads to or deepens political conflict, as in the case of Darfur, the recent civil war-torn region of Sudan, contributing to the breakdown of ordered rule (i.e. **good governance**) and thus providing seedbeds for forces opposing the onward march of global integration. At another level, China's energy requirements are growing to fuel its investments in infrastructure (more than 300 million people are expected to migrate to the urban areas in the coming two decades), its manufacturing sector and its commitment to private forms of transport. This is is leading to growing competition for access to oil and gas in Africa and other resource-rich regions, with as yet unknown consequences for the global political architecture (see also Chapter 2). Will it herald a new phase of imperialist rivalry for resources, eerily similar to that of the late nineteenth century?

What is to be done?

We are accustomed to think positively, to find a solution for every problem (Figure 15.1). This is just as well, since the threats confronting humankind in general, and the historically advantaged west in particular, are very substantial. I cannot pretend that I have an answer to the growing problems besetting the globalizing economy, but there are three environment-related issues which need to be addressed.

First and foremost is the environmental challenge in general, and climate change in particular. The aggregate numbers are overwhelming—the biosphere simply cannot withstand the pressures that sustained global growth will place on it. This has multiple consequences. We need to find a more efficient path for the generation, distribution, and consumption of energy, developing a range of energy-saving technologies and organizational structures. More rational—that is higher—energy-pricing is one part of the solution, but it is only one. If it is the only solution, then higher energy prices will exacerbate the gap between the global haves and the global have-nots, both within and between countries. We also need to reduce material consumption patterns in the rich countries, placing greater emphasis on leisure and services, particularly if consumption in the developing world is going to grow as living standards are raised. China adds a new

Figure 15.1 Sorting out globalization? G8 world leaders in Tokyo, Japan, 2008.

coal-powered power-station every four days (see Chapter 2), and however efficient new forms of carbon capture might be, the expansion of energy consumption in the emerging economies is only environmentally sustainable if consumption in high income countries is reduced.

Second, for many people and many countries, globalization is less a route to higher living standards and more a force for immizeration (economic impoverishment) (Kaplinsky 2005). The injunction to deepening globalization involves a fallacy of composition—that is, it works for individual countries, acting in isolation, but not when all countries follow the same route (Mayer 2002). Unfettered access to global competition, in the context of global excess capacity, means impoverishment for those without unique skills. This is the case for much of Africa and Latin America, and for a growing number of people in western Europe and North America (whose reserve armies of labour are not just China and India, but also eastern Europe and central America).

In principle, incomes can be provided for these marginalized communities through the redistribution of the fruits of increased growth and productivity, but realpolitik stands in the way—the rich will not pay the taxes required to fund this redistribution, and they are increasingly able to avoid taxes in a world of liberalized financial flows. I therefore believe that we need to revisit the virtues of protection, but in a world of regional preferences which will at the same time allow for enhanced scale and productivity growth with a less extended and energy-intensive transport infrastructure. This means a greater emphasis on intraregional integration reflected in trade, governance systems, financial flows, migration, and other elements which have become so important in the globalizing world of the past few decades.

Third, the entry of China (and soon India) as a growing source of global economic demand and supply needs to be reflected in the global political architecture. For decades this has been the playground of the rich Western countries. For example, the head of the World Bank has always been an American nominee, whilst that of the International Monetary Fund is a European appointee. How durable this division of spoils turns out, will be an indicator of the degree to which currently hegemonic powers are willing to accommodate new entrants to the global political table.

These three factors interact. The politics of growing inequality, economic imbalances between newly emergent Asian economies and the US and the EU, and unemployment are leading to growing opposition to globalization in many countries who have previously enthusiastically embraced it. To an increasing extent, opponents of globalization are focusing on its environmental excesses as a key point of opposition, particularly as energy prices rise. In hindsight, seen from a perspective of the earth's biosphere, the unequalizing tendencies inherent in current structures of globalized production systems may prove to have been a godsend.

■ SUMMARY

- There is a widespread 'triumphalist' belief that globalization as we know it is unstoppable, but three issues challenge this belief:

- The nature of capitalist development leads to the degradation of the global commons, to global warming and to climate chaos. This is because the search for profitability leads to over-consumption by many, and because of the extended and energy-intensive transport arteries of global production systems.

- Current processes of globalization deepen inequality. This leads to growing political hostility to globalization from the marginalized, and deepens poverty, often reinforcing pressures on the environment.

- The combination of environmental impacts and inequality sets up internal contradictions which are likely to undermine the sustainability of the globalization process itself.

All of the above requires us to think about new and innovative ways in which humankind can continue to survive and prosper, living with, rather than in command of nature.

■ REFERENCES

Bernstein, J. (ed.) (1990) *The Food Question: Profit vs People*. London: Earthscan.

Cornia, A. C. and Court, J. (2001) *Inequality, Growth and Poverty in the Era of Liberalization and Globalization, Policy Brief No. 4*. Helsinki: Wider.

Dew-Becker, I. and Gordon, R. J. (2005) *Where did the Productivity Growth Go? Inflation Dynamics and the Distribution of Income, Working Paper, 11842*. Cambridge, MA: National Bureau of Economic Research. Available from http://www.nber.org/papers/w11842.

Groves, A. S. (1996) *Only the Paranoid Survive*. New York: Doubleday.

Hira, R. (2004) *Implications of offshore sourcing*. Mimeo, Rochester Institute of Techonlogy.

Kaplinsky, R. (2005) *Globalization, Poverty and Inequality: Between a Rock and a Hard Place*. Cambridge: Polity Press.

Marx, K. (1876/1970) *Capital: A Critique of Political Economy, vol. 1*. London: Lawrence and Wishart.

Mayer, A. J. (2002) The fallacy of composition: A review of the literature. *The World Economy* 25(6): 875–894.

Milanovic, B. (2005) *World Apart: Measuring International and Global Inequality*. Princeton, NJ: Princeton University Press.

Ricardo, D. (1817) *The Principles of Political Economy and Taxation*. London: Dent (reprinted 1973).

Sachs, W. (2004) Climate change and human rights. *Development* 47(1): 42–49.

Schumpeter, J. (1942) *Capitalism, Socialism and Democracy*. London: Unwin University Books.

Smith, A. (1776/1976) *An Enquiry into the Nature and Cause of The Wealth of Nations*, 4th edn. Oxford: Oxford University Press (reprinted 1976).

Stern, N. (2006) *The Economics of Climate Change: The Stern Review*. Cambridge: Cambridge University Press.

Toynbee, P. (2003) *Hard Work. Life in Low Pay Britain*. London: Bloomsbury.

Wikipedia. http://en.wikipedia.org/wiki/Main_Page, accessed 11 January 2009.

Williamson, J. G. (1998) Globalization, labor markets and policy backlash in the past. *Journal of Economic Perspectives* 12(4): 51–72.

World Bank (2002) *Globalization, Growth, and Poverty: Building an Inclusive World Economy*. Washington, DC: World Bank and Oxford: Oxford University Press.

World Bank (2007) http://siteresources.worldbank.org/DATASTATISTICS/Resources/reg_wdi.pdf, accessed 17 September 2007.

■ **FURTHER READING**

Dicken, P. (2007) *Mapping the Changing Contours of the Global Economy*. London: Sage.

Milanovic, B. (2005) *World Apart: Measuring International and Global Inequality.* Princeton, NJ: Princeton University Press.

Kaplinsky, R. (2005) *Globalisation, Poverty and Inequality: Between a Rock and a Hard Place*. Cambridge: Polity Press.

Stern, N. (2007) *The Economics of Climate Change: The Stern Review*. Cambridge: Cambridge University Press.

■ **USEFUL WEBSITES**

http://www.unctad.org. The annual Trade and Development Report and the World Investment Survey are excellent sources of data and heterodox views on globalization.

http://www.asiandrivers.open.ac.uk. A resource site which focuses on the impact of China and India on the global economy in general, and on developing economies in particular.

http://www.globalvaluechains.org/. Focuses on the operations of global value chains.

16

Environmental ethics and development

Martin Reynolds

Introduction

> *More than any other time in history, mankind faces a crossroads. One path leads to despair and utter hopelessness. The other, to total extinction. Let us pray we have the wisdom to choose correctly.*

> *Woody Allen, American humorist, quoted in Westley et al. (2006: 90)*

Humour often provides respite in a perceived world of intractable dilemmas. Local issues such as access to clean water or availability of food can be driven by, as well as contribute to, global issues such as climate change and the global economy. Take for example the issues around constructing large-scale dams. The Narmada Dam Project in India (Figure 16.1) is one of the longest-standing development and environmental controversies of its kind. Box 16.1 summarizes the history and some key issues. These issues are complex and generate questions of responsibility.

The conflicts are formidable. Large-scale dam construction, like other big socio-economic developments such as air travel expansion, has been subject to criticism, both through extensive consultant reporting and strong activism and protest. Yet often there is a sense of inevitability about such projects. Decisions appear to be made through some inescapable march of so-called progress. So perhaps Woody Allen is right to be cynical, but cynicism belies a wealth of opportunities for seeing and doing things differently.

An ethical outlook on such issues can help to realize such opportunities. For example, looking behind Woody Allen's acerbic observation, some basic ethical questions might be asked to reveal areas of responsibility that need to be and can be managed more constructively.

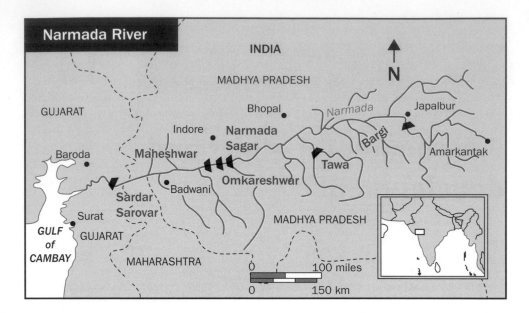

Figure 16.1 Map of the Narmada Valley and proposed dams.

BOX 16.1 THE NARMADA DAM PROJECT

The Narmada Dam Project in India involves the construction of 30 large, 135 medium, and 3,000 small dams. These will exploit the waters of the river Narmada and its tributaries for improved irrigated agriculture to produce more food, and the generation of hydroelectric power. The idea was first conceived in the 1940s by India's first Prime Minister, Jawaharlal Nehru, but it was not until 1979 that the project took form. Of the 30 large dams, Sardar Sarovar is the largest and most controversial. In 1979, the Sardar Sarovar Project was proposed and attracted initial support from international financial institutions including the World Bank. However, after much controversy and protest, particularly since the late 1980s, many financial institutions withdrew support. Protest was led by Narmada Bachao Andolan (NBA), a national coalition movement including people affected by the project, environmental and human rights activists, scientists, and academics.

   The construction of Sardar Sarovar dam itself was stopped in the mid-1990s. However, in October 2000, the Indian Supreme Court gave a go-ahead again for the construction of the dam. Other dams associated with the wider Narmada project have likewise been developing, come under criticism, and have been the subject of protest.

Four general issues can be summarized in relation to the Narmada Project:
- Water access and quality (e.g. water-borne diseases from stagnant reservoir waters).
- Urban and rural economic development (e.g. displaced populations from rural areas).
- Change in agricultural practices (e.g. shift towards large-scale irrigated farming).
- Ecological impacts (e.g. loss of biodiversity in previously rich hydrological systems).

Source: Friends of River Narmada (2008)

1. What are the particular issues that need attention? Does global warming deserve more attention than longer standing issues of abject poverty in the world? Or should we just despair at the magnitude and complexity of issues confronting us?

2. How might these issues be attended to and by whom? Is it just 'them' out there or is it also you/me/us? Or should we just resort to fatalism, nurturing a general sense of apathy and blame?

3. Why are some issues privileged more than others, and some ways of dealing with them prioritized over others? What opportunities are there for challenging main-stream ways of dealing with harmfulness and wrongdoing? Or should we just remain cynical of human nature and the prospects to realize alternative ways of doing things?

Despair, apathy and cynicism are all too prevalent in modern society. Moreover they are human attributes sometimes encouraged by those with an interest in keeping things as they are—contributing to vicious cycles of business as usual and the type of eco-social collapse invoked by cynics. In what follows I'll use each of the three sets of questions above in turn to explain how ideas from environmental ethics can help guide more purposeful engagement with environment and development dilemmas. The Narmada Dam Project is used to ground the discussion.

Despair? Issues and values: normative ethics

Ethics concern contrasting questions of 'is' with questions of 'ought'. This is sometimes referred to as normative ethics. The 'is' comprises a descriptive world of issues that are experienced by different stakeholders. The 'ought' comprises a normative world of values, often multiple and conflicting, which are used by stakeholders to make judgements on the realities they experience. Many issues relating to environment and development are experienced as complex questions requiring continual attention to value judgements on what ought to be.

So what role do value judgements play? Arguments for and against the Narmada Project can be considered as expressions of value judgements: arguments for construction based on judgements on what ought to be the outcome, and arguments against based on judgements regarding what ought not to be the outcome (Box 16.2).

One of the key problems arising from any controversial issue is sorting out judgements of fact from value judgements. Scientific information on levels of domestic water supply, power generation, agricultural production, estimated numbers of poor and underprivileged communities being dispossessed of their livelihood, and ecological impacts, are vital. However, professed levels of impact, both positive and negative, are often contested even amongst scientists. 'Fact' and value are inextricably linked. So being aware of accompanying value judgements is also very important. Ethics makes values explicit. Box 16.3 provides an understanding of different types of value and different perspectives.

Arguments in support of the Narmada Dam Project can be said to have an anthropocentric perspective with a dominant, instrumental value judgement on water as a resource. Few would deny this as an important value judgement, particularly in a context of poor access to clean

BOX 16.2 NARMADA DAM PROJECT : VALUE JUDGEMENTS

Those arguing for construction: the dam should be built because it will deliver the following outcomes which we judge to be good/right/valuable

Those arguing against construction: the dam should not be built because it will deliver the following outcomes which we judge to be bad/wrong/worthless

1. Water access and quality

- Supply water to 30 m people including drinking water facilities
- Irrigate crops to feed another 20 m people covering 17,920 km² of land.

- Increase prospect of insect-borne diseases.
- Inundate areas causing salinization of land alongside canals through build up of salts.

2. Urban and rural economic development

- Provide hydroelectric power
- Improve access to electricity in remote villages
- Develop facilities for sophisticated communication systems in the project areas
- Increase employment both in construction and post-construction maintenance

- Dispossess large numbers of poor and underprivileged communities of their land as a source of livelihood
- Provide inadequate compensation and rehabilitation for resettled people as with previous experiences in India
- Over-estimate power generated and under-estimate likely long-term dependence on private transnational companies
- Prompt excessive profiteering amongst private contractors and possible corruption in dispensing large budgets

3. Agricultural practice and technological development

- Modernize agricultural practices using irrigated farming
- Provide irrigation infrastructure for biofuel agricultural production (and genetically modified crops)
- Develop fisheries industry

- Lose skills in more sustainable farming practices
- Undermine confidence in experts
- Give false promises regarding maintenance of dams given seemingly disorganized State infrastructure
- Disrupt downstream fisheries

4. Ecological impacts

- Protect against advancement of desert and provision of flood protection to riverine reaches
- Establish wildlife sanctuaries protecting rare species (e.g. sloth bear, wild ass, bustard)

- Diminish biodiversity through monoculture irrigated farming
- Devastate existing riverine ecosystem
- Submerge current forest farmland
- Ignore possible long-term impacts (e.g. large reservoirs could cause earthquakes)

BOX 16.3 VALUES AND PERSPECTIVES ON ENVIRONMENT AND DEVELOPMENT

Values are an assessment or measure of the worth of something.

Two broad types of value can be distinguished in environmental ethics:

1. *Instrumental* is the value that something has as a means to an end. So money might be good *only* because it leads to other good things (purchase of goods). Putting monetary value on environmental goods, or considering nature in terms of natural resources, and ecosystem services, are typical expressions of instrumental value in relation to nature.

2. *Intrinsic* is the value that a thing has in itself, or for its own sake, or in its own right. Money for example is not intrinsically good (unless you are a collector of historic or different currencies) whereas most other goods might arguably be considered as having some intrinsic value. Environmentalism as a social movement in the mid-twentieth century grew from an appreciation of the intrinsic value for nature.

A third type of value can be associated with the valuer as against the valued. Here, value is linked with obligations and the boundaries of the *moral community*—who or what is worthy of respect (past, present, future generations? other animals? all living organisms? ecosystems? biosphere? universe? multiverse?).

3. *Personal* (or individual) is the internally held value of the valuer usually attached to character traits such as having integrity. Behind any value is a valuer with particular perspectives on the world guided by personal values. Two perspectives on the environment based on personal values can be distinguished—anthropocentric and ecocentric.

An *anthropocentric* perspective is one that places humans in a privileged position over nature. An extreme position of anthropocentrism—egocentrism—privileges individual humans. Other extremes assume that the destiny of humanity is to conquer and master the forces of nature. Such a perspective assumes that nature is only valuable insofar as humans have a use for it, in terms of human needs (i.e. instrumental valuation).

An *ecocentric* perspective is one that regards human beings as simply one part of a moral community consisting of all living things as well as non-living natural objects (rivers, mountains etc.). Humans no longer occupy a privileged position at the top of the moral community.

water. From a more ecocentric perspective, claims are made of providing flood protection for ecosystems, and offering compensation to support sanctuaries for endangered species.

Anthropocentric arguments can also be made against the project. The displacement of communities, loss of livelihood, and diminished access to water amongst vulnerable groups are particularly significant. The possible loss of biodiversity through deforestation and increased salinization will have aesthetic disadvantages which can also be factored in from an anthropocentric perspective. Many of the arguments against such projects, however, derive from a more ecocentric perspective, bringing attention to wider and longer term ecological impacts.

However, values and perspectives are not fixed entities. They vary and develop according to the context and time in which they are applied. This is evident with the Narmada case study. As time has moved on, protest around the Narmada has become symbolic of a global concern for how we engage with nature and the long-term consequences. Environmental ethics helps to explain such changes in terms of different types of value judgements and perspectives. Environmental ethics therefore help to make sense of arguments for and against a project, and to respond effectively to such arguments using the appropriate language of value and perspective. In short, rather than despair at the complexity of issues arising, environmental ethics provides a handle—a vocabulary around value judgements—for appreciating and dealing with issues more constructively.

So with a means of surfacing value judgements, what guidance might be given towards using those judgements for responsible action?

Apathy? How to do ethics and be ethical: philosophical ethics

Whilst normative ethics helps in revealing the interplay of value judgements, more specific questions on what to do can draw on traditions of moral philosophy. *Philosophical* ethics is about searching for answers to questions about:

1. Doing what's good (or harmful).
2. Doing what's right (or wrong).

The first question invites consideration of the consequence of a decision and appropriate ways of measuring the consequences. The second invites consideration to the *intention* behind any decision and any particular obligations behind such intention. Table 16.1 provides some ideas about the kinds of benefit/harm, and rights/wrongs that might be looked at in association with each of the four main issues arising from the Narmada Dam Project. (It should be noted that, as with any philosophical abstraction, the categorization into 'good' or 'right' is a slightly artificial one and there is not always a clear distinction between them.)

The responses to each question—what's good and what's right—in relation to any issue can be contested. For example, on the issue of agricultural practice some might suggest that a more appropriate 'good' from an anthropocentric perspective would be to improve

Table 16.1 Philosophical ethics: what to do?

Issues around the Narmada Dam Project	Doing what's good (not harmful) *Measures of success*	Doing what's right (not wrong) *Intentions and obligations*
Water access and quality	Improve quality of water and access to clean water (avoid disease and drought)	Provide universal access to clean water (not reinforcing or developing skewed, unequal access)
Urban and rural economic development	Improve quality of life for citizens (avoid poverty and use of only economic indices)	Provide opportunity for all humans to flourish (not constraining humans from flourishing)
Agricultural practice and technological development	Improve range of productive capacities for farming (avoid loss of ecologically sustainable farming skills)	Provide expertise to support appropriate practice (not contriving a simplistic solution)
Ecological impacts	Improve quality of the natural environment (avoid ecological deterioration)	Provide protection against ecological destruction (not ignoring wider obligations to nature)

intensity of production. Further contestation may arise in privileging one type of question over another. Should 'rights' and obligations be advanced in spite of the effects of action, or vice versa? An obligation to respect nature may for example be inappropriate in circumstances where the effect is to further human impoverishment. Similarly, a focus on maximizing human welfare may infringe on the rights of other life-forms to flourish. Reference to value judgements and associated perspectives (Box 16.2) can help make sense of these conflicts.

Environmental ethics also addresses character attributes around being ethical or environmentally responsible. This invokes a third tradition in philosophical ethics drawing upon Western (e.g. Ancient Greek) and Eastern (e.g. Buddhism and Taoism) philosophy:

3. Being virtuous (or non-virtuous).

Table 16.2 summarizes some virtues and non-virtuous (vice) character attributes that might be associated with each of the Narmada issues.

Table 16.2 Philosophical ethics: how to be?

Issues around the Narmada Dam Project	Virtue	Vice
Water access and quality	Justice	Injustice
Urban and rural economic development	Moderation	Greed
Agricultural practice and technology progress	Humility	Arrogance
Ecological impacts	Compassion	Recklessness

Identifying human character attributes can help to ground instinctive feelings about responsibilities and make more relevant discussion about consequences of good action and the obligations implied by right action. As well as mediating discussion between Western and Eastern traditions, it also helps with appreciating the many helpful ideas of human nature relationships amongst existing indigenous tribal communities.

The risk of philosophical ethics though is in confining it to academic discourse—a sure way of generating cynicism! So another task is to keep ethics alive and integral to all deliberations around planning and politics.

Cynicism? Ethics and politics

Ethics is, by its very nature, contested terrain. Disparate value judgements and perspectives, contested ideas on what is good and what is right, and indeed contested virtues (should justice have privilege over compassion? is compassion an appropriate environmental virtue?), all clearly need appropriate space for deliberation. When reviewing the list of four issues associated with Narmada you may personally feel aggrieved at the priority given to an anthropocentric perspective on the issues (ecological impacts being the last and only issue that privileges nature). What opportunities exist for countering value judgements and the development of alternative viewpoints? In other words, what political space exists to openly challenge assumptions underpinning development initiatives? Political space, by which I mean opportunities for actors to manoeuvre and put their case on an issue, most commonly though not exclusively through demonstration (Figure 16.2), discussion, and debate in both formal and informal settings, represents the interface between ethics and politics. An engagement with environmental ethics demands political space to avoid being seen as the cosy arena of armchair philosophy prompted by cynics. Box 16.4 illustrates examples of such space in India.

It would be foolish to pretend that groups with more radicalized anthropocentric and ecocentric perspectives have 'won' the battle in Narmada against conventionally dominant economic interests. However, it would also be misleading to underestimate the political space nurtured through the engagement of environmental ethics with social and political theory, policy advisors, and activist groupings.

An attitude of cynicism can undermine attempts at fostering new political space. Vandana Shiva provides a helpful riposte:

> *The big transformations always seem to move in the direction of destruction. But if you look at the small actions, the hundreds of people saying 'I will speak against human rights violations, I will be part of the voice', at the thousands of farmers who work with us who have created an alternative agriculture in spite of the dominant policy, that's where change is happening, and that change will continue to grow.*

> *Vandana Shiva in Davis (2008: 29)*

Shiva raises three important virtues: *hope* in countering the despair of real world violations; *purposefulness* in countering an understandable apathy of a farming

---

BOX 16.4 ETHICS AND OPENING POLITICAL SPACE IN NARMADA

- *Narmada Bachao Andolan (NBA)*: their non-violent campaigns, including hunger strikes, a 36-day march, mass demonstrations, and use of the media, prompted institutions as powerful as the World Bank to withdraw support from the Project in the early 1990s.
- *Friends of River Narmada (FRN)*: an international coalition of individuals and organizations (primarily of Indian descent) supporting NBA in terms of providing a repository of information, ongoing research, public education and outreach, promotion, and publicity.
- *Navdanya movement*: participatory research initiative set up in 1991 to counter corporate control over farming practices. Though not directly related to Narmada dams, Navdanya nurtures practices other than monocrop industrial agriculture promoted as part of large-scale dam projects.

Conventionally political space is dominated by men. Women activists such as NBA spokesperson, Medha Patkar, and Vandana Shiva, founder of Navdanya, belong to a tradition of ecofeminism reflecting an important gender imperative for creating new political space.

Sources: Friends of River Narmada (2008), Navdanya (2008)

Figure 16.2 The massive rally in Khandwa, Madhya Pradesh on 4 June 2007 involving people affected by the Omkareshwar and Indra Sagar dams.

community subjected to industrialized agricultural policy; and *trust* in countering the cynicism that change from business as usual is unattainable due to human nature.

▓ SUMMARY

- Environmental ethics can provide support towards seemingly intractable questions of environmental responsibility that otherwise can lead to despair, apathy, and cynicism.
- An understanding of normative values and perspectives—normative ethics—can help surmount a sense of despair.
- Practice in thinking about doing what's good, doing what's right, and being responsible—philosophical ethics—can help to overcome apathy.
- Cynicism needs to be continually checked through creating space for engaging more passionately with normative and philosophical ethics.
- Each of the three domains of environmental ethics—normative, philosophical, and political—can be aligned with more constructive virtuous attributes, for example, attributes of hopefulness, purposefulness, and trust.
- Any one environmental situation being addressed whether personal, local, national or global, requires attention to all three domains of environmental ethics.
- Environmental ethics alone is not *the* answer, but it can provide precious support in guiding and keeping alive the right questions.

■ REFERENCES

Davis, R. (2008) Making waves: interview with Vandana Shiva. *New Internationalist* (410): 29.

Friends of River Narmada (2008) http://www.narmada.org/introduction.html, accessed 25 April 2008.

Navdanya (2008) http://www.navdanya.org/about/index.htm, accessed 4 August 2008.

Westley, F., Zimmerman, B. and Patton, M. Q. (2006) *Getting to Maybe*. Canada: Random House.

■ FURTHER READING

Attfield, R. (2003) *Environmental Ethics: An Overview for the 21st Century*. Cambridge: Polity.

Des Jardins, J. R. (2001) *Environmental Ethics: An Introduction to Environmental Philosophy*. Belmont, CA: Wadsworth.

Light, A. and Rolston, H. III (eds) (2003) *Environmental Ethics: An Anthology*. Malden (USA), Oxford (UK), Victoria (Australia), Berlin (Germany): Blackwell Publishers.

Reynolds, M., Blackmore, C. and Smith, M. J. (eds) (2009) *The Environmental Responsibility Reader*. Milton Keynes: The Open University and London: Zed Books. Contains a version of this chapter amongst 31 readings bringing together issues of environmental ethics with systems thinking, social learning, and political studies in ecological citizenship.

■ USEFUL WEBSITES

There are many sites providing in-depth analytical support on questions of environmental responsibility. Two sites helpful in providing research support towards issues of environmental justice are:

The Transnational Institute: **http://www.tni.org/detail_page.phtml?=&text10=news_ctw-news&menu=11c**

The Corner House: **http://www.thecornerhouse.org.uk/index.shtml.** The Corner House carries out research and advocacy with the aim of linking issues for informed discussion and strategic thought on critical environmental and social concerns. Both sites provide open-access documentation on a wide range of ethical issues relating to environment and development.

Another important source for current information on climate change legislation is the Intergovernmental Panel on Climate Change: **http://www.ipcc.ch/**

There are an increasing number of very influential environmental advocacy groups with a strong online presence providing helpful information on ethical issues associated with the environment. Some are more country and/or issue focused such as the Navdanya and Friends of River Narmada sites given in the references above, or the Indonesian group Down to Earth: http://dte.gn.apc.org/. Others are more international, such as Greenpeace: **http://www.green-peace.org/international/**

Climate Justice Programme: **http://www.climatelaw.org/** under the auspices of Friends of the Earth. A collaboration of lawyers and campaigners around the world providing up-to-date information for encouraging, supporting, and tracking enforcement of the law dealing with issues of climate change.

Environmental ethics online directory: **http://gadfly.igc.org/ee-list.htm** provides a list of names and contact details of scholars and professionals working in the field of environmental ethics.

Section C review

Section B introduced conceptions of environment, development, and sustainability based on place, whether a single city, a country, or a region. In so doing, it also introduced several of the pressing environmental issues facing today's world—climate change, biodiversity, conservation, water, and waste. The chapters of Section C have elaborated on these major themes. Inevitably the analysis has often been related to impacts on place and places, but these pressing 'big' environmental and related issues, which invariably know no borders, form the main focal points of our enquiry. In this way, the chapters of Section C are a different, but equally instructive way of cutting the environment, development, and sustainability cake.

The issues themselves are quite distinctive, with their own concepts and scientific ideas for understanding them. Each forms the basis, therefore, of a separate chapter but they are also interconnected and so our key themes of difference and interdependence carry through from Section B. This section therefore reinforces the first proposition from Chapter 1, that environment and development are inextricably linked. Some of the chapters also touch on the scientific knowledge and principles that underpin and explain these links.

One of the major stories to have gripped the world in recent years has been that of human-induced climate change through carbon emissions. Although a global phenomenon, it has local impacts in different parts of the world, of which floods, droughts, and rising sea levels are among the three most quoted. In Chapter 10 Roger Blackmore explains how our understanding of what is happening in climate change and why it is happening has been illuminated by an interdisciplinary collaboration of scientists and other academics under the auspices of the Intergovernmental Panel on Climate Change (IPCC).

An international consensus in terms of scientific understanding is a necessary precondition for action, but it is not necessarily sufficient. The causal chain linking human economies with climate change:

economic growth $\longrightarrow$ energy consumption $\longrightarrow$ carbon emissions $\longrightarrow$ climate change

comprises one of the core tensions of environment, development, and sustainability and is hard to break.

However, Blackmore points out that development of 'clean technologies', such as zero-carbon renewable energy sources can make a big difference and be a win–win opportunity, but inevitably they require investment. The paradox is that many of the poorest countries of the world will be hardest hit by climate change, yet they hold most of the world's renewable energy resources while lacking the capital to develop them. It will be in everybody's interest to assist the required investment. As with the earlier chapter in Section B on China, the emphasis in Chapter 10 is on a reworking of the economics.

A major impact of climate change is expected to be on biodiversity, which is the focus of Michael Gillman in Chapter 11. In the context of tropical rainforests, his chapter empha-sizes the importance of maintaining biodiversity to human livelihoods, whether local or at scale. Original ideas about sustainability referred to optimum harvesting levels of spe-cies, such as fish, by humans so that they could be renewed. Gillman also engages with the view held by many who see the solution to both the world's environmental problems and enduring poverty in small, self-reliant communities living in harmony with their environment. An oft-quoted example is of forest dwellers trading with the outside world in renewable forest products, such as fruit, leaves, and medicinal derivatives. One coun-ter-argument is that such a 'solution' relies on low population densities if these products are not to be quickly exhausted (and hence not renewed). Another concerns the validity of the poverty–biodiversity loss linkage and Gillman quotes studies which show no link between reduced poverty and reduced deforestation.

Related to biodiversity, but more explicitly place-specific, is the older notion of conser-vation. This is the focus of Roger Wheater and Adeline Muheebwa in Chapter 12. While the two authors write about very different places—respectively the Highlands of Scotland and the tropical forests of Uganda—conservation is again seen to be in tension with economic activity. The actual modes of economic activity are, however, quite different in the two contexts. In Scotland, at least in part, it concerns the visitor/tourist spin-offs from preserving a dramatic landscape. Echoing Gillman, in Uganda it is about different forms of livelihoods—those which depend on sustainable use of forest products, and those which would clear away the forests for large-scale commercial activities (in this case a sugar cane plantation). Wheater and Muheebwa also each address the conflicts among multiple interests when attempting to manage the tensions. In Scotland there are well-worn institutional mechanisms which, at least in theory, enable broad representation of different interests in resolving conflicts. In Uganda such mechanisms are as yet under-developed and conflict can take on violent forms.

Relevant to the approach adopted in these three chapters is the scientific background of three of the authors. Roger Blackmore (Chapter 10) has a physics, Michael Gillman (Chapter 11) an ecology, and Roger Wheater an ecology/zoology background. Gillman and Wheater both point to the importance of rigorous and replicable experimentation, even when this takes place in the 'real life' of the field where, unlike laboratory experi-ments, one cannot control all the variables. Gillman, especially, uses scientific studies as the evidence base of his chapter. Blackmore too starts tellingly from the sound sci-entific base of the Intergovernmental Panel on Climate Change (IPCC). The IPCC has authoritatively told us what the problem is, why it is happening, and what is necessary, in scientific terms, to deal with it.

We should also note that these three chapters are connected by the same principle derived from environmental science. This is the carbon cycle which describes the interac-tion between carbon and oxygen, on which life on Earth is based. As Chapter 10 makes clear, the levels of carbon dioxide in the Earth's atmosphere are increasing because of human disruption to the equilibrium of this interaction. At the centre of the causal chain leading to climate change that is described above, is the link between energy consumption and carbon emission. As humans burn fuels to meet our energy needs, carbon dioxide is released in accordance with this simple equation:

$$carbon + oxygen \longrightarrow energy\ release + carbon\ dioxide$$

Thus human exploitation of energy resources such as oil and coal is accompanied by the release of a pollutant that contributes to global warming.

This same equation, in reverse, forms the 'other side' of the carbon cycle and is the reason why the continued existence of forests, as described in Chapters 11 and 12, is so vital.

$$carbon\ dioxide + energy \longrightarrow carbon + oxygen$$

This reversed equation describes, in its most basic form, the process of photosynthesis. Green plants remove carbon dioxide from the air and, using the energy of sunlight, produce oxygen, on which nearly all life on Earth depends.

Water is another issue where conservation in its biophysical sense is often applied as a concept, although the resource is of such fundamental importance to all life that it is rightly seen as an issue in its own right. It is also the focus of another principle of environmental science, the water or hydrological cycle which lies at the heart of Chapter 13 by Sandrine Simon. The hydrological cycle concerns the global circulation of water between the oceans, the atmosphere (i.e. cloud formation), and land. Humans typically meet their needs for water by extraction from rivers, lakes, and groundwater, or, in other words, tapping into the water cycle at the point between precipitation on to land and discharge into the ocean. Not only does this affect the quantity of the water but also, when the used water is returned, its quality through the introduction of pollutants. As described by Simon, organic pollutants from sewage, agriculture, and some industrial sources, contributed to turning the Rhine into 'Europe's cesspit'. The Rhine was able to recover partly because these discharges have been controlled but also because of natural processes within river water where micro-organisms break down organic matter into relatively harmless substances.

Water is not surprisingly a resource over which there is often conflict, which is the main focus of Chapter 13. This can be between countries, where rivers form borders (and hence are shared) or cross from one country to another on their route to the sea. It can also be in-country, where activity upstream affects the river downstream. Conflict can further occur over multiple uses in a single location, or across locations. Through a discussion of mechanisms for resolving international conflicts along the Rhine in Europe since 1815, Simon develops the important point suggested by Wheater and Muhweeba that institutional mechanisms for resolving environmental conflicts evolve only through long periods of time.

Pre-eminent of the many uses for water is its sustenance of human life, and hence its public health dimensions in terms of its availability and quality. Public health is an issue everywhere, but particularly in cities due to the high density of human populations (Section B, Chapter 9). Another major issue for public health in cities is the quantity of waste they generate and waste management forms a further major environmental theme for Stephen Burnley and Richard Kimbowa (Chapter 14).

These authors contrast waste management policies in two large cities—Birmingham in the UK and Kampala in Uganda. In Kampala, the focus of policy is indeed on maintaining and enhancing public health. This fundamental purpose of waste management is considered to have been met in relatively affluent Birmingham, however, where policy now focuses on typical Western European concerns (highlighted in Section B, Chapter 5) of recycling and recovery. An interesting conclusion to this chapter suggests that the relative affluence of Birmingham and other northern cities might produce greater gains for public health globally if some of their financial resources were transferred to improve waste management in poorer southern cities.

Cities are hubs of economic activity with attendant environmental pressures. In the penultimate chapter (15) of the section, Raphael Kaplinsky takes the relationship between environment and economy to the global scale and considers the enormous pressures on the environment through the extension and deepening of economic linkages internationally, which we know as globalization. Kaplinsky makes the further point, however, that the structure of economic growth should also be seen in terms of relational inequality, meaning that it benefits a small proportion while the majority suffer from increasing job insecurity (including among the professional classes) if not impoverishment. This would be a depressing analysis were it not for the possibility that the combination of environmental pressures and the conflicts arising from deepening inequality could well draw a halt to the globalization process itself. Globalization might not be so inevitable, as its proponents claim, after all.

Inequality and conflict are issues that frequently raise questions of our fundamental values. Finally in this section, Chapter 16 by Martin Reynolds on environmental ethics is about disentangling and making explicit the relationship between facts and values. The chapter illustrates how decisions are rarely clear-cut because the stakeholders have multiple perspectives and value positions. For example, how do you compare the desirability of a big dam project in India (or elsewhere) which proponents claim will serve a national good, with the project's undoubtedly adverse impacts on local people who literally have to be moved out of the way in order for it to proceed?

Also running through this section is the second proposition from Chapter 1 that groups aspects of sustainability into three areas of concern, namely environment, society, and economy. These three have all been represented in the chapters of Section C. The first two chapters could be categorized as related to the environment and its science basis; society, particularly in the form of institutions, is evident in Chapters 12, 13, and 16; economics is clearly a focus for Chapters 14 and 15, but to categorize in this way would be to overlook the interdependence of these areas. Running through the book we see, therefore, an interdependence of people and places, and also between two fundamental issues—environmental change and economic activity. In addition, we have described briefly an interdependence *within* the biophysical world, which is inherent in the carbon and hydrological cycles. Scientific principles such as these are an underlying foundation of the whole book but they are especially evident in this section where the Earth's systems and cycles are a feature of many chapters.

If human activity is disturbing the equilibrium in these cycles, what might a different kind of human activity do about it? The chapters in Section D look at actions generally across different scales which attempt to make sense of and formalize the complex interdependencies of this section as forces for the better.

SECTION D
Action for environment, development, and sustainability

17

Local governance and sustainable development in Wales: making a difference?

Alan Thomas

Introduction

Wales is unusual in having a commitment to sustainable development (SD) written in to its constitution. Within the United Kingdom, Wales has enjoyed devolved government since the Government of Wales Act 1998 established the National Assembly for Wales (NAW or Welsh Assembly). Section 121 of this Act places a statutory duty on the NAW to promote sustainable development in all its business.

This chapter looks at the **governance** of SD in Wales and asks what difference this statutory duty has made. What has actually happened as a result?

The empirical parts of this chapter mostly come from a study undertaken in 2003-04 (Williams and Thomas 2004). Despite changes since then, the examples still illustrate general concepts and principles.

The context: the architecture of governance for sustainable development in Wales

Wales (Cymru in Welsh) is part of the UK. Situated to the west of England, its population was almost 3 million in 2006, out of just over 60 million for the UK as a whole. Wales has a distinctive culture, including rugby as its national sport, and its own ancient language, spoken by just over 20 per cent of the population. Its south-east is post-industrial with a long tradition of working-class solidarity, whereas mid, north, and west Wales are sparsely populated and rural.

Wales' small size may be an important factor in its governance. Relatively few agencies are involved even in complex issues and networks are relatively close-knit and overlapping, with key individuals playing several roles. According to one senior manager, 'The beauty of Wales is its manageability in terms of scale—there is a real chance to take SD forward' (quoted in Williams and Thomas, 2004: 4).

Welsh devolution brought no power to raise taxes or make laws—although more recently the Welsh Assembly has gained limited legislative competence. The main freedom is in how to spend the annual block grant from the British government, received mostly for health and education, plus aspects of housing, culture, agriculture, and transport. In some cases distinctive policies have been implemented compared to England.

Like other complex policy areas, SD is not amenable to 'policy as prescription', in which problem analysis shows what measures should be undertaken by government to lead linearly to solutions. Rather, the promotion of SD should be seen as a process of steering and influencing involving public action by a number of agencies (see Wuyts *et al.* 1992). This is why this chapter is about **governance**—'the sum of the many ways institutions, public and private, manage their common affairs' (Commission on Global Governance 1995: 2)—rather than government. Theorists of governance such as Rhodes (1996) and Stoker (1998) argue that government does not have a monopoly of power or authority, so that governance involves networks of state, private, and non-governmental actors.

For SD in Wales, these networks focus on the Welsh Assembly Government (WAG). They include British government agencies, because UK laws and regulations still cover Wales, and also European and global networks in which WAG has been active. Within Wales the main agencies are:

- Local government: 22 local authorities (and 22 Local Health Boards with the same boundaries and a joint duty to prepare Health, Social Care, and Well-being Strategies)
- The Welsh arms of national and international NGOs working in environment and development, such as:
  - the Royal Society for the Protection of Birds (RSPB): the largest UK conservation organization in terms of membership
  - World Wildlife Fund (WWF) -Cymru and Oxfam-Cymru, which help promote global linkages whilst acting locally.
- Welsh NGOs:
  - those involved in anti-poverty work or conservation locally
  - the national umbrella Welsh Council for Voluntary Action.
  - Cynnal Cymru (The Sustainable Development Forum for Wales), an important new NGO formed as the voice of Welsh **civil society** on sustainable development
- Assembly Sponsored Public Bodies (ASPBs): the Welsh quangos such as the Countryside Council for Wales, Environment Agency, and Welsh Development Agency.

A combination of NGOs, quangos, and other interested parties such as academics played a major lobbying role in the run up to devolution. Thanks to their efforts the Welsh Assembly not only has to promote SD but do so in an open and inclusive way (Flynn *et al.* 2003). Thus, WAG promotes particular policy processes which engage with multiple stakeholders from both inside and outside government from the formative stages of policy development. These include:

- Accountability and funding mechanisms between the Assembly and the ASPBs.
- Formal partnership arrangements set up with different sectors including voluntary, private, and local government.
- Community consultation and regeneration.
- Regulation mechanisms for public service in Wales through performance management.
- Sharing best practice at different levels.

Thus, although it was not deliberately designed as a coherent system, Wales has its own particular institutional and governance framework for managing SD.

What would 'success' look like?

Many protagonists claim that there is now a general consensus over what is meant by SD, and that the problem is about implementation. They invariably quote the Brundtland definition (see Chapter 1 of this volume) and the idea of integrating social, economic and environmental considerations into all policy making (Williams and Thomas 2004: 8). Certainly, within WAG, SD is treated as a known 'thing', a future desired state, progress towards which tends to be interpreted in terms of indicators and measurement. This fits well with the prevailing paradigm of performance management.

In evaluating the first SD scheme, the Assembly adopted a small suite of headline indicators (Table 17.1), most of which were similar to those used at UK level. Although other indicators have been added, this set is still in use.

However, this approach fails to represent the trade-offs or even outright conflicts in SD decision making. Which indicators carry most weight? What if a proposed policy to increase employment is also likely to increase greenhouse gas emissions? Asking such questions makes it clear that, despite the claims for consensus, a huge variety of meanings is ascribed to the notion of SD, reflecting values conflicts and philosophical differences (see Table 17.2). These range from 'weak' conceptions of SD as a distinct policy area connected with green or environmental issues, through those which see SD in terms of integration, effectively equating it with good joined-up decision making, to 'strong' conceptions of SD as a philosophy or set of principles to be applied to the activities of all policy sectors.

The question then arises whether SD stands above other general concerns or becomes just one of a list together with equality, social inclusion, and the Welsh language, all of which provide principles which WAG applies to all policy areas. If SD does everything then everything may be regarded as aspects of SD, so that it may lose the specificity of its future-orientation and global principles.

Authors such as Jacobs (2006) maintain that SD conforms to Gallie's (1956) notion of an 'essentially contested concept'. In other words, part of the essence of the concept is that, even with a measure of agreement on principles, there can be no single correct interpretation.

Table 17.1 Sustainable development indicators for Wales.

Policy area	Indicator
Employment	% of people of working age in work
Education	% of people at age 19 with at least an NVQ (National Vocational Qualifications, which are UK work-related, competence-based qualifications set at different levels) level 2 or equivalent
Crime	Crime rates per 100,000 population for: • theft of and from a vehicle • burglary in a dwelling • violent crime
Housing	% of unfit dwellings
Climate change	Emissions of greenhouse gases: basket of greenhouse gases
Air quality	Days per year when air pollution is moderate or high: • urban; Cardiff, Swansea, Port Talbot • rural; Aston Hill, Narberth
River water quality	% of river lengths of good or fair quality: • chemical quality • biological quality
Wildlife	Population of wild birds
Waste	Household waste and amount recycled: • total household waste • household waste recycled or composted
Welsh language	% of people who are Welsh speakers: • all aged 3+ • children aged 3–14
Electricity from renewable sources	% of electricity produced in Wales generated from renewable sources
Ecological footprint	Wales' global ecological footprint in area units per person

Source: Welsh Assembly Government (2002).

Political differences and clashes of interests lead to inevitable differences in usage of such concepts. For example, the Welsh Development Agency interprets the combination of economic, social, and environmental to include economic growth as part of SD. However, others see the principle about environmental carrying capacity as requiring a limit on economic growth.

Table 17.2 Meanings of sustainable development.

Sustainable development as . . .	
Contested framework	Framework encouraging discussion around the process and content of sustainable development
Overarching principles or philosophy	Limited carrying capacity of the natural environment Local–global links Intergenerational equity Avoiding irreversibility
Policy integration	Integrating social, economic, and environmental considerations/impacts Working in partnership Empowering people and communities Evidence-based and outcome focused Medium term time frames
Policy area	Policy area concerned with the natural environment, including energy, waste, biodiversity

Source: Williams and Thomas (2004: 8).

Interestingly, Gallie also suggested that disputation over the meaning of an essentially contested concept may contribute positively to the ideals from which it is derived. This might imply that there has been progress towards SD if in order to justify a course of action one has to argue about the meaning of SD, alongside traditional concerns such as costs, standards, or efficiency.

Thus SD is perhaps best conceived as a contested framework: 'Invoking sustainable development principles cannot settle policy dilemmas, but it can offer an overarching framework to facilitate debate around areas of tension and disagreement' (Williams and Thomas 2004: 12).

Making a difference: ways of achieving change

The different interpretations of SD suggest different types of action. Seeing SD as a policy area leads to the design and implementation of specific SD policies or projects apart from policies in other areas. Williams and Thomas (2004) call this policy separation. By contrast, interpreting SD to mean policy integration implies mainstreaming SD principles by incorporating them into integration frameworks throughout an organization or more widely. We look at each of these, as well as a third type of action aimed at achieving institutional change.

Specific SD projects or policies (policy separation)

There are many examples at both national and local levels of promoting SD through projects or policies in particular sectors. At its simplest this is about the management of environmental issues such as waste management, energy efficiency, and biodiversity within specialist departments. It works best when there is a clear mandate and finance allocated to provide a service, such as, for example, collection of recyclable waste by local authorities.

This approach also includes projects and policies that incorporate SD principles in any sector, not only those directly concerned with environmental issues. Almost all Welsh local authorities have one or more dedicated SD practitioners, who have developed a range of projects around food, schools, and health as well as energy and recycling (Netherwood 2003). These may opportunistically develop into cross-departmental initiatives or involve linkages between several organizations. However, extremely limited SD budgets keep such actions small-scale, fragmented, and dependent on external funding.

Mainstreaming SD principles (policy integration)

Various frameworks and mechanisms already apply to the activities of, say, a local authority as a whole, in which SD principles could be made prominent. Here are some examples of integration opportunities that could significantly impact on SD outcomes. They do not make an exhaustive list and are not mutually exclusive.

Community strategies

Local authorities have a statutory obligation to prepare community strategies which bring together their whole range of activities with those of other agencies in the locality—public, private, and voluntary—to achieve strategic aims for the community, including a contribution to SD. The NAW circulates advice on preparing these (National Assembly for Wales 2001), which incorporates principles that chime well with SD such as partnership working, community participation, and linking social, economic, and environmental considerations within a single framework.

Community strategies are often conceived in a matrix framework with four or five policy areas and SD as one of a number of cross-cutting themes, such as community safety, Welsh language, equality, and social inclusion. SD does not have priority, and there is little consensus on its meaning, so the potential for promoting SD may be lost.

Wales Programme for Improvement

Local authorities in Wales are obliged to undertake self-assessment using a performance management framework called the Wales Programme for Improvement (WPI). The checklist of aspects to be assessed includes 'the arrangements for making sustainability... integral to its work', but no more specific advice is given.

Williams and Thomas (2004: 24) note mixed results for SD the first time this process was undertaken. In one authority, the SD function was highlighted as high risk, resulting in its being moved to a more central organizational position in the Chief Executive's department, and an SD training and awareness programme was introduced for elected members and key officers. In another review: 'We only looked at SD in terms of our own housekeeping—paper recycling—not whether it was reflected in our services' (quoted in Williams and Thomas 2004: 24). Although this sounds weak, such reviews are subject to audit, which may challenge the situation—assuming that bodies such as the Audit Commission in Wales have a 'strong' version of SD incorporated into their own performance management framework.

Spatial planning

Spatial planning, 'the consideration of what can and should happen where' (Welsh Assembly Government 2004: 3), has potential to contribute to SD policies and outcomes which is increasingly acknowledged. All local authority Unitary Development Plans are subject to a sustainability appraisal, and a Good Practice Guide from the Assembly sets out how this can best be undertaken. At national level, the Wales Spatial Plan (Welsh Assembly Government 2004) is clearly seen as incorporating SD principles: The section headed 'Purpose' emphasizes that 'we place the core values of sustainable development in everything we do' (Welsh Assembly Government 2004: 4).

Promoting institutional change

Commitment to SD principles is not fully agreed throughout the many organizations involved. Specific SD projects often coexist with 'business as usual' in other departments, while introducing SD into organization-wide processes like the WPI above can lead to a grudging 'tick-box' compliance which leaves established ways of thinking about, say, economic development, intact.

Institutions can be organizations or established aspects of wider society which embody common values and ways of behaving. An example of the former is the UK National Health Service with its value of free health care; an example of the latter is the institution of marriage. *Institutional* change concerns changing these values and ways of behaving. It describes what is required to spread SD principles throughout organizations and embed them into management practices in all sectors. This means change not only in organizational structures, processes, goals, and mission statements, but also in the mindset which determines informal ways of working, behaving, and justifying decisions.

Working towards such change generally means influencing rather than commanding resources to achieve goals directly. It goes well beyond simply putting certain tasks or projects into practice, although there are strong reasons for linking with practical implementation activities.

This 'influencing' can be done both from outside and within particular organizations. A good example of influencing from outside was the lobbying done by NGOs and others which led to the relatively strong duty towards SD being placed on the NAW at devolution. More recent examples include the activities of organizations such as Cynnal Cymru and Sustainable Development Co-ordinators Cymru (SDCC). The latter is a national professional network (Marsh and Rhodes 1992) formed by SD practitioners in Welsh local government (and the three Welsh National Parks). SDCC has had some success in pushing for SD principles to have greater prominence in Wales-wide integrative processes such as the WPI and Wales Spatial Plan.

SD coordinators in local authorities have also attempted to act as internal change agents. This requires actions to be found and projects designed which promote the values of SD at the same time as achieving useful results.

If one is looking for integrated adoption of SD principles across the whole of an organization's programmes, such projects may appear insignificant and fragmented. However, they may be justified as exemplars, demonstration projects, or 'policy experiments' (Edwards and Hulme 1992; Rondinelli 1993). As noted in Williams and Thomas (2004), this can work in two ways. Within a particular sector, a successful pilot project can demonstrate good practice. It may then be duplicated, modified as a result of learning from evaluating the original, and eventually used to build new policy for the whole sector.

The Sustainable Schools award scheme provides an exemplar in this sense. This was devised and designed by two SD practitioners in a Welsh county council, to encourage the integration of sustainable development into all aspects of school management as well as curriculum. By 2003, twenty schools had signed up to the scheme. Various supporting organizations, including environmental NGOs and a National Park Authority, provide advice, resources, and teacher training.

Not only has this scheme influenced this particular council, which now has an official policy for Education for Sustainable Development and Global Citizenship (ESDGC), and fed back into other local community issues such as safe routes to schools and debates on school transport budgets, but it has also been influential at the national level. WAG has adopted a national strategy and produced a guidance document on ESDGC, which has also been integrated into the inspection regime of Estyn, the Welsh Schools Inspectorate. ESDGC is one of the cross-cutting themes in the new Welsh schools curriculum and the influence of Sustainable Schools is evident in the continued promotion of an integrated approach linking school management with curriculum.

The other way is to improve the public visibility of the issue and to make it less abstract. An example was a project undertaken by a SD practitioner in a local council to make one of Wales' most prestigious rugby clubs carbon-neutral for one season. This was done not with any expectation of it becoming a long-term policy beyond that season or of 'scaling up' to other rugby clubs, but to show what was possible, create opportunities for education and for political credit, and hence increase both organizational and public support for SD in the area.

One of the strongest elements of institutional change, therefore, is that it aims to change values (see also Chapter 16). This creates a number of problems or issues, particularly around responsibility and accountability.

Where does the authority to promote value change come from? SD may have been adopted as a policy commitment on paper, but the corresponding values are certainly not embedded throughout a local authority, a typical school, or WAG itself. And exactly what are the values which should underlie SD? This is itself not agreed, not surprisingly since the principles of SD are still a matter for contestation, and likely to continue so.

In terms of accountability, it has been well argued that focusing on performance targets and measurements can actually be inimical to sustainable development (e.g. Fowler 1997). The spread of values and principles may be as important as implementing and measuring specific projects or tasks—but this does not make it easier to demonstrate accountability.

Conclusion: has Wales' statutory duty made a difference?

- Have the measurements of the headline indicators improved?
- Are there more projects aimed at implementing SD policies in different sectors?
- Are the various integration mechanisms more embedded and do they incorporate SD principles in such a way as to give them prominence and promote practices in line with them?
- Has there been institutional change towards embedding SD values into key agencies?

The answer to all these questions is: to some extent, yes. Arguably, however, the extent and pace of change has not been sufficient. Much more radical change is required to meet the challenge of sustainable development. Wales' statutory duty has provided some opportunities and stimuli, but it is not yet clear if there has been more progress compared to how things would have been without it.

■ SUMMARY

- Wales has a statutory commitment to sustainable development (SD) and its own particular institutional and governance framework for managing SD.
- SD may be regarded as an 'essentially contested concept' with no single correct interpretation.
- SD is variously treated as a desired future state, a distinct green policy area, equivalent to integrated policy making, and a set of principles to incorporate into all policy areas.

- These different interpretations of SD give rise to different types of action: the design and implementation of specific projects and policies (policy separation); building SD principles into organization-wide frameworks (policy integration); and spreading commitment to SD principles by changing organizational values, e.g. through the use of exemplars by change agents (promoting institutional change).

- Although there are examples of all these types of action in Wales, it is unclear if the statutory duty towards SD is making a positive difference.

■ ACKNOWLEDGEMENT

Much of the empirical material in this chapter comes from a 2003-04 study financed by the Joseph Rowntree Foundation, undertaken in a team led by Paul Williams, then of the National Centre for Public Policy at Swansea University.

■ REFERENCES

Commission on Global Governance (1995) *Our Global Neighbourhood: The Report of the Commission on Global Governance*. Oxford: Oxford University Press.

Edwards, M. and Hulme, D. (1992) *Making a Difference: NGOs and Development in a Changing World*. London: Earthscan.

Flynn, A., Netherwood, A. and Bishop, K. (2003) *Multi-level Governance and Sustainable Development in Wales*. Cardiff: Centre for Business Relationships, Accountability, Sustainability and Society (BRASS), Cardiff University.

Fowler, A. (1997) Assessing development impact and organizational performance. In *Striking a Balance: A Guide to Enhancing the Effectiveness of Non-Governmental Organisations in International Development*, pp. 160–183. London: Earthscan.

Gallie, W. B. (1956) Essentially contested concepts. *Proceedings of the Aristotelian Society* 56: 167–198.

Jacobs, M. (2006) Sustainable development as a contested concept. In A. Dobson, *Fairness and Futurity: Essays on Environmental Sustainability and Social Justice*, pp. 21–45. Oxford: Oxford University Press (reproduced from *Oxford Scholarship Online Monographs*, April 1999).

Marsh, D. and Rhodes, R. A. W. (1992) *Policy Networks in British Government*. Oxford: Clarendon Press

National Assembly for Wales (2001) *Preparing Community Strategies*. Cardiff: National Assembly for Wales.

Netherwood, A. (2003) *A Survey of Sustainable Development Activity within Local Authorities and National Park Authorities in Wales*. Cardiff: Cardiff University.

Rhodes, R. A. W. (1996) The new governance: governing without government. *Political Studies* 33: 652–667.

Rondinelli, D. (1993) *Development Projects as Policy Experiments*. London: Routledge.

Stoker, G. (1998) Governance as theory: five propositions. *International Social Science Journal* 155(March): 17–28.

Welsh Assembly Government (2002) *Sustainable Development Indicators for Wales*. Cardiff: Welsh Assembly Government.

Welsh Assembly Government (2004) *People-Places-Futures: The Wales Spatial Plan*. Cardiff: Welsh Assembly Government.

Williams, P. and Thomas, A. (2004) *Sustainable Development in Wales: Understanding Effective Governance*. York: Joseph Rowntree Foundation.

Wuyts, M., Mackintosh, M. and Hewitt, T. (eds) (1992) *Development Policy and Public Action*. Oxford: Oxford University Press.

18

Cities and climate change: leading to a low carbon London

Godfrey Boyle

Introduction

At some time during the early years of the twenty-first century it is estimated that, for the first time in human history, the majority of the world's population will be living in cities. If this trend towards increasing urbanization continues, as seems likely, it is estimated that by 2050 approximately 70 per cent of the human population will be city-based (United Nations Population Division 2007). This shift from rural to urban living implies that solutions to the problems of environmental and social sustainability will increasingly need to focus on cities (see also *Science* 2008; Starke 2007).

London is the capital city of the United Kingdom and one of the world's leading business, financial, and cultural centres. The population of Greater London in 2006 was 7.5 million, making it the most populous municipality in the European Union.

The challenges to humanity of climate change caused by human-induced emissions of greenhouse gases, principally carbon dioxide, have been well described elsewhere (for example, Chapter 10 this volume; IPCC 2007; Stern 2007). The principal effects of climate change on London and similar major world cities, if mitigation and adaptation measures are not adopted, are likely to include increased summer temperatures leading to premature deaths among vulnerable populations, increased extremes of rainfall leading to excess surface run-off of water and local flooding, and rising sea levels necessitating improved defences against flooding of low-lying areas.

Carbon emissions in London

Residents of cities, like the residents of non-urban settlements, cause climate change due to emissions of carbon dioxide (CO_2) and other greenhouse gases through their use of carbon-based fossil fuels. However, energy use in cities is often more efficient than in rural areas.

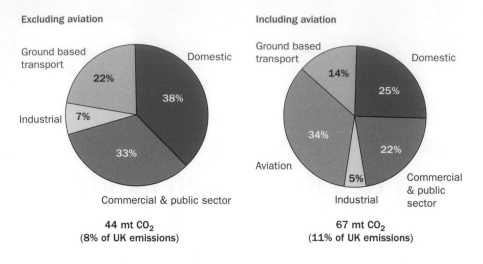

Figure 18.1 London CO_2 emissions in 2006 by sector, excluding and including aviation.

Housing density is higher, leading to lower energy per unit of floor area, transport distances are shorter, and there is better provision of public transport, which is more energy-efficient than private vehicle use, etc.

Cities, on the other hand, are often vulnerable to climate change. Even in the absence of greenhouse gas-induced temperature rises, the urban heat-island effect leads to higher ambient temperatures in cities than in non-urban locations; and major cities are often located near the sea, so they are vulnerable to sea level rises caused by global warming.

During the 1990s, London's carbon dioxide emissions from fossil fuel use reduced slightly, from about 45 million tonnes of CO_2 in 1991 to about 42 million tonnes in 1999. They then rose slightly to around 44 million tonnes in 2006. The distribution of emissions between sectors is shown in Figure 18.1.

Reducing London's carbon emissions

In the *London Mayor's Energy Strategy* (Greater London Authority 2004), the Greater London Authority (GLA) proposed ambitious plans for major reductions in London's carbon emissions. Central to the strategy was the deployment of decentralized energy systems, in the form of high-efficiency combined heating/cooling and power generation (CCHP) systems and renewable energy sources. CCHP involves capturing the 'waste' heat from electricity generating systems and putting it to use in heating and/or cooling buildings.

One key aim of the strategy was for London to double its installed CCHP capacity by 2010, compared with 2000. The strategy also envisaged the deployment by 2010 of 25,000 solar water heating systems, 7,000 solar photovoltaic (i.e. electricity-generating) roofs, 500 small wind turbines, and six large wind turbines. All London boroughs (the subunits of

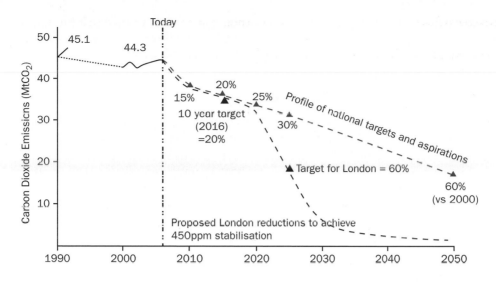

Figure 18.2 Profile of projected London carbon emissions 1990–2050 compared with UK national Government projections.

Source: Greater London Authority (2007) *Action Today to Protect Tomorrow. The London Mayor's Climate Change Action Plan.*

local government in London) would have at least one zero-carbon development (i.e. with zero net carbon emissions, measured annually) and were urged to implement targets for renewable energy. All major developments over 1000 m² in floor area would need to be approved by the Mayor. To gain approval they would need to incorporate energy/carbon-saving measures such as passive solar design and natural ventilation, and should utilize CCHP and renewable energy wherever possible. The aim was for renewables to supply enough electricity for 100,000 homes and heat for 10,000 homes by 2010. These targets were to be tripled by 2020.

In 2006 the energy consultancy PB Power produced a four-scenario study for the GLA, *Powering London into the 21st Century* (PB Power Energy Services Division 2006). This showed that London, through vigorous deployment of CCHP and other decentralized energy sources, could reduce its CO_2 emissions by about 20 per cent by 2020.

In addition, the GLA supported ambitious plans to site several major offshore wind farms in the estuary of the River Thames, downstream from London—although this area is just outside the authority's jurisdiction. These include the 1000 MW London array, the 500 MW Greater Gabbard array and the 300 MW Thanet array. These are scheduled for implementation around 2011 and should supply the domestic electricity needs of about a quarter of London households.

The reports and initiatives mentioned above laid the groundwork for the *Mayor's Climate Change Action Plan*, published in 2007 (GLA 2007). The Plan's main aim was ambitious: to achieve a 20 per cent reduction in London's CO_2 emissions by 2016 and a 60 per cent reduction by 2026 (see Figure 18.2). This aim is more challenging than the UK

Government's target of 60 per cent by 2050 – though it should be noted that this target was subsequently revised upwards to 80 per cent by the Government's Climate Change Committee, set up as part of the Climate Change Act 2008.

Some key aspects of the Action Plan

One of the key measures proposed by the GLA to mitigate emissions of greenhouse gases, principally CO_2, included the establishment of a London Climate Change Agency (LCCA) to implement the strategy. The LCCA in turn set up a London Energy Supply Company (ESCO), in partnership with the major electricity utility EDF Energy, to roll out CCHP and other decentralized energy schemes across the capital during the coming decade and beyond.

Another key measure is the use of the Mayor's planning powers of scrutiny over all major new building developments. These are being used to reduce CO_2 emissions from new buildings. They involve a mandatory three-stage approach to reducing energy use and CO_2 emissions, summarized in the slogan 'Be Lean, Be Clean, and Be Green'.

- Be Lean: use less energy. This involves implementing stringent energy conservation measures.
- Be Clean: use energy efficiently. This involves using combined heat/cooling and power (CCHP) and other energy-efficient technologies wherever possible.
- Be Green: use renewable energy. This involves supplying a significant proportion (10–20 per cent) of the energy demand from renewable sources.

CO_2 savings in the domestic, commercial, and public sectors

In the domestic sector, the Action Plan measures include establishing a Green Homes Programme, including an energy and greenhouse emissions awareness campaign, a one-stop advice service, and making available subsidized insulation for home owners, particularly those on lower incomes. For householders who can afford to pay, a Green Concierge Service has been set up, offering energy and emissions audits, including recommended measures to reduce emissions, and a project management service to facilitate their implementation.

For the commercial and public sectors the LCCA has established a Green Organisations Programme. This includes a Better Buildings Partnership between the LCCA and major building owners, and a Green 500 badging scheme for organizations, to encourage emission-reducing behaviour changes and improved operations by building occupants and managers.

CO_2 savings in the energy supply and new buildings sectors

In the energy supply sector, the Action Plan measures include encouragement for major increases in the use of combined heat and power (CHP) and, where appropriate, combined cooling, heat, and power (CCHP). It also encourages generation of energy from waste; the use of on-site renewable energy sources, such as passive solar space heating, solar water heating, solar photovoltaics, biofuels, ground-source heat pumps, building-integrated wind turbines, and large-scale renewable energy projects located near to London.

Some renewable energy sources briefly explained:

- passive solar space heating: design of buildings so that energy from the sun is captured and retained to warm the interior, e.g. by using glass roofs in parts of the building.

- solar water heating: using energy from the sun to heat water, e.g. a circulating system in which water passes behind a transparent panel that faces the sun.

- solar photovoltaics: solar panels that convert energy from the sun into electricity.

- biofuels: fuels derived from biological material, usually plants or animal wastes, excluding fossil fuels which are derived from long-dead biological material.

- ground-source heat pumps: a system that takes advantage of heat retention in the soil and extracts it via a network of buried pipes.

New buildings in London are only a small proportion of the total building stock, most of which is old. For new buildings the Action Plan proposes revisions to the London Plan to encourage further use of decentralized energy, and improvements in advice and support to building developers and to the staff in the 32 London boroughs responsible for building control and for granting planning permission for smaller developments. It also proposes funding exemplary low- and zero-carbon developments by the London Development Agency (LDA) such as the Gallions Park development, the 2012 Olympic Village, and projects in the London boroughs such as the Beddington Zero Energy Development, 'BedZed', in Croydon

Progress in greening new developments in London

To what extent has the Mayor's scrutiny of planning applications for major developments been successful? Southbank University in South London (Day *et al.* 2009) studied a sample of 131 such planning applications, out of 350 made in 2004–06. Their main conclusion was that GLA policies are likely to lead to savings of about 1 million tonnes of CO_2/year by 2010 in new buildings. Approximately a quarter of these savings are from renewables; the rest is from CCHP and improved energy efficiency. The renewables contribution was small when the policy was first introduced, but a 10 per cent contribution is now routinely achieved and the renewables target is to be raised to 20 per cent.

Achieving CO_2 savings in ground-based transport

Measures in the Action Plan to mitigate emissions from ground-based transport, including cars, lorries, buses, taxis, rail, and the London Underground system, include major investments to improve the efficiency of the public transport system, more efficient vehicle operation, promoting low carbon vehicles and fuels, and introducing carbon pricing for transport.

Prior to the publication of the Action Plan, various measures had already been introduced to cut London's transport emissions. These included the introduction of congestion charging, involving a payment by motorists of £8 a day (2009) to drive into the city centre. Low- and zero-emission vehicles are exempt from the charge. As a result, traffic levels in central London have decreased by about 15 per cent. The proceeds of the congestion charge have been used to fund a shift to public transport involving many more and newer buses, improvements to the rail and Underground network, and better facilities for cycling and walking.

Other transport-related measures include requiring London Underground (which uses about 3 per cent of the city's electricity) to buy power from renewable energy suppliers, and introducing a Low Emission Zone in which it is mandatory for large vehicles entering the Greater London area to meet strict emission standards. The GLA also aims to convert all 8000 London buses to lower-emission 'hybrids' (with combined diesel-electric engines) and is supporting the London Hydrogen Partnership, which has introduced three zero-emission fuel cell/hydrogen-powered buses into regular service.

The bigger picture: the C40 Cities Climate Leadership Group

In addition to assuming a leading role within the UK, London has taken a lead in encouraging similar developments in other major world cities. In 2005 London convened the first meeting of the C40, a forum of 40 of the world's largest cities concerned about climate change. At a subsequent meeting in New York in May 2007, the mayors of the C40 cities joined with the Clinton Climate Initiative, led by ex-US President Bill Clinton, to announce the creation of the global Energy Efficiency Building Retrofit Programme. This brings together four of the world's largest Energy Service Companies (ESCOs), five of the world's largest banks and, initially, sixteen of the largest cities. It offers building owners an energy audit to quantify current energy use and building emissions, recommendations on a full range of measures to reduce emissions, a comprehensive, discounted offer of goods and services guaranteed to deliver the identified reductions, and an optional financing element to pay for the works, paid back through the guaranteed energy savings.

In late February 2008 it was announced that two major energy service companies had won the first contracts to implement energy- and carbon-saving measures in London's public buildings. In the words of the Mayor's announcement:

> *World leading energy service companies Dalkia and Honeywell have been selected to help cut energy use in Greater London Authority buildings by 25 per cent and the Mayor will now launch a procurement process to let every public sector organization in London benefit from the same deal. London is the first city in the world to have completed the municipal tender process and appointed a company following a deal developed by the Clinton Climate Initiative for the C40 group of cities . . . The Mayor will also be working with the Clinton Climate Initiative to sign up private sector businesses to green their buildings.*

> *GLA (2008a)*

The role of political leadership

In developing and implementing London's ambitious plans to mitigate climate change, the role of strong political leadership has been crucial. From the beginning of his term of office in 2000, London's first Mayor, Ken Livingstone, recognized the seriousness of the climate change problem, the need for London to take action, and the need for further supporting action at national and international governmental levels. As he later wrote in his foreword to the *London Climate Change Action Plan*:

> It is almost impossible to exaggerate the danger of climate change. I have no doubt that it is the single biggest threat to the future development of human civilisation … The aim of this Plan is to deliver decisive action in London with the urgency that is required.

> The Action Plan demonstrates that London can make deep and meaningful cuts in its emissions through actions by London public authorities, by businesses, and by individual Londoners. For the next ten years we can meet the target reductions that scientists say are necessary . . . The difficult truth, however, is that without action at national and international level we cannot continue to achieve this through to 2025 and beyond.

> *Livingstone (2007: iii)*

Inevitably, the processes of democracy and consequent changes to the political party in power will potentially lead to a lack of policy continuity. When Boris Johnson was elected Mayor of London in May 2008 there was some uncertainty that this climate change policy leadership would continue. However, the initial indications are that the overall climate change policy goals will be retained, although the strategy for achieving them may change. As Johnson commented in a press announcement to the C40 cities in June 2008:

> I am delighted to be working with the Mayor of Toronto and the C40 cities in recognition of my commitment for London to play a significant role in reducing carbon emissions, and tackle climate change. London is a world leader in this area, and I am determined to see that the expertise and innovation being developed here is shared with cities around the world. Equally, I see many excellent ideas coming from others' cities that we want to learn much more about—I have already committed to introduce a public bike hire scheme in London similar to that developed in Paris, another C40 city. Cities across the world share the common threat of climate change, and cities create most of the carbon emissions that are causing it, so it is vital we continue to work together to accelerate action on this issue.

> *GLA (2008b)*

Conclusions

If enacted in full, the measures in the London Climate Change Action Plan should enable a reduction in CO_2 emissions of about 27 per cent by 2025, but in the GLA's view this is not enough—a reduction of at least 60 per cent by 2025 is needed if London is to contribute its share of the global CO_2 reductions needed to stabilize climate. To get to

60 per cent, additional national Government action is required, to set a firm price for carbon emissions, and to remove regulatory obstacles to deploying CCHP and renewables.

Progress so far has been modest, though a reduction of one million tonnes in CO_2 emissions from new buildings is likely to have been achieved by 2010. It is probably too early to judge the success of the GLA's plans—especially if they are subject to significant changes under the new post-2008 political leadership.

In addition to possible political changes, however, further problems in implementing CO_2 reductions in London include the very large scale of the investment required—hundreds of millions of pounds per year. This represents a formidable challenge, especially in view of the 'credit crunch' that (at the time of writing) is inhibiting investment in many large-scale projects around the world. National, and possibly international, government action may be required to galvanize public and private investment on the scale that is required.

■ SUMMARY

- This chapter investigates the issues of climate change with respect to London as an exemplar of a major world city. It describes:

  - the plans of the Greater London Authority (GLA) to mitigate climate change by reducing emissions of carbon dioxide;

  - the main measures by which the GLA intends to achieve these planned reductions;

  - how successful these measures have been to date; and the extent to which London's climate change initiatives have depended on political leadership.

- It concludes that modest progress in emission reductions has probably been achieved in the short term, but that further, longer-term progress will depend on the continuation of strong political leadership and on the availability of capital for investment in the various measures required.

■ REFERENCES

Day, A. R., Ogumka, P., Jones, P. G. and Dunsdon, A. (2009) Use of the planning system to encourage low carbon energy technologies in buildings. *Renewable Energy* doi: 10.1016/j. renene.2009.02.03.

Greater London Authority (2004) *Green Light to Clean Power: The Mayor's Energy Strategy*. London: Greater London Authority.

Greater London Authority (2007) *Action Today to Protect Tomorrow: The Mayor's Climate Change Action Plan*. London: Greater London Authority.

Greater London Authority (2008a) Mayor announces start of a groundbreaking programme to green London's public buildings. Press release, 28 February, available from http://www.london. gov.uk/view_press_release.jsp?releaseid=16634.

Greater London Authority (2008b) Mayor of London pledges commitment to international change co-operation. Press release, 28 June, available from http://www.london.gov.uk/view_ press_release.jsp?releaseid=17194.

IPCC (Intergovernmental Panel on Climate Change) (2007) Summary for Policy Makers. In S. Solomon, D. Qin, M. Manning, Z. Chen, M. Marquis, K. B. Averyt, M. Tignor and H. L. Miller (eds), *Climate Change 2007: The Physical Science Basis. Contribution of Working Group I to the Fourth Assessment Report of the Intergovernmental Panel on Climate Change*. Cambridge and New York: Cambridge University Press.

Livingstone, K. (2007) Foreword. In *Action Today to Protect Tomorrow: The Mayor's Climate Change Action Plan*. London: Greater London Authority.

PB Power Energy Services Division (2006) *Powering London into the 21st Century. Report for the Mayor of London and Greenpeace*. London: PB Power.

Science (2008) Reimagining Cities: special issue, 8 February 2008.

Starke, L. (ed.) 2007) *State of the World: Our Urban Future. Worldwatch Institute Report*. New York and London: W.W Norton.

Stern, N. (2007) *The Economics of Climate Change: The Stern Review*. Cambridge: Cambridge University Press.

United Nations Population Division (2007) *World Urbanisation Prospects: The 2007 Revision Population Database*. New York: United Nations. Available from http://esa.un.org/unup/, accessed June 2008.

■ USEFUL WEBSITES

Greater London Authority: http://www.london.gov.uk

London Climate Change Agency: http://www.lcca.co.uk

London Energy Partnership: http://www.lep.org.uk

C40 Climate Cities: http://www.c40cities.org

19

Exploring participation in science, technology, and innovation: tissue culture bananas in Kenya

Joanna Chataway, Peter Robbins, and James Smith

Introduction

Science, technology and **innovation** are key to 'good change' (Chambers 1997) in developing countries. They are therefore major areas of investment for development agencies and the subject of numerous agricultural and environmental initiatives. Yet they are also often contentious. Two key issues are who should make decisions about new technological innovation and whether more or less participation is needed in decision making processes.

Participation for the purposes of this chapter relates to the inclusion of stakeholder(s) in decision making processes. It tends to be viewed in two ways:

- As a mechanism to ensure increased democracy in decision making about development.
- As a way of achieving social and economic development goals, such as economic growth or better ways of delivering health and education.

The former approach represents the 'entitlement view' and the latter the 'productivity view' (Mosse 2005) of participation. Both views are widely used and promoted.

Entitlement in the sense used in this chapter is associated with our rights, such as the right of stakeholders to influence decisions that are likely to have an impact on them.
Productivity is associated with effectiveness of decision making processes, for example the claim that inclusive participation of a range of stakeholders will lead to better decisions and better enactment of those decisions.

Participation is drawn into debates about science and technology (S & T) in many ways and in various settings. The quality and purpose of participation is often contentious. In particular, discussions about efforts to create forms of direct or participative democracy in relation to science and technology have been bitterly contested (Irwin and Wynne 2004; Leach *et al.* 2005; Jasanoff 2007). Recent work highlights problems in using 'participation' as a decision making tool about technologies in the abstract. Various authors argue strongly against advocating the use of participative techniques simply on the grounds that they lead to increased democracy because sometimes such processes can be hijacked by special interest groups (Tait 2004; Taverne 2005).

Other work highlights instances where formal 'participation' is used to legitimate S&T-based development projects and in communication to funders who require evidence of broad stakeholder involvement, yet it has not delivered significant opportunities for wider engagement over the way those projects are developed and run (Chataway and Smith 2006). Participation events do not necessarily represent significant advances in deliberative democracy; that is where the participants are able to interact in open discussion with each other and, out of their differences, advance a collective knowledge which informs their decisions and activities. They often fail, therefore, in terms of both productivity and entitlement. This chapter also suggests, however, that it is fruitful to think about how participation can contribute to building common values and ways of behaving—known as **institutional development**—in the area of S&T innovation rather than as one-off events, smokescreens, or vehicles for legitimation.

We explore these issues at a theoretical level first, through revisiting Albert Hirschman's framework of 'exit, voice and loyalty'. We then use an example of the development of tissue culture bananas in Kenya to argue that building institutions that embody new principles of institutional development is often complex and difficult and subject to numerous pressures.

> Albert Hirschman (1915–) is a German-American economist, who first became known for his work in development economics. His (1958) book *Strategy of economic development* developed the notion of 'unbalanced growth', arguing that developing countries lack entrepreneurial activity rather than capital, which is curtailed by institutional weaknesses. As such, he looked at how non-market forces shape market activity.

Participation and engagement: creating 'voice'

Ideally the 'entitlement view' and the 'productivity view' of participation become one and the same. In the short run, more participation, more inclusion and more accountability lead to improved performance, 'better development' and 'better' projects. In the longer term it lays the basis for better decision making and a more solid grounding for development. These can and do happen, but of course it does not always happen. Entitlement and productivity as a result of participation can happen independently of each other or not at all.

Development, as a process of change, has conflict at its heart. One way of looking at participation is as a channel for voicing dissent—it can be seen as a way of ensuring engagement even in the case of disagreement (Hirschman 1970). Hirschman's work on 'exit', 'voice' and 'loyalty' draws on economics and politics and is an explicit attempt to create an understanding of how effective **institutions**, which embody common values and ways of behaving, get created.

Economists tend to value exit, the ability to choose and reject options to buy products and services, as a way of promoting competition and better institutions. However, not all situations lend themselves to this market based institutional mechanism and Hirschman posits the more political mechanisms of voice and engagement as an equally important way to advance institutional development. Simply put, the opportunity to participate, to engage, gives people the opportunity to voice their opinions and, given the right institutional conditions coupled with *loyalty* to the endeavour, participation can bring improvement. A balance between opportunities for voice, possibilities for exit, and efforts to secure loyalty are fundamentals in creating effective organizations.

Hirschman maintains that the process of creating conditions for engagement and voice involve both the productivity and entitlement aspects of participation. His thoughts are echoed in more recent work in innovation, understood broadly as new or evolving knowledge which is put to productive use (Chataway 2005). For example, the idea of feedback between researchers, producers, users, and consumers is one articulation of voice and participation in the innovation systems literature (Lundvall 1992; Hall 2005).

Such feedback between the varied stakeholders is often relatively unstructured and informal. It is different from having structured, organized, boundaried forms of participation which are promoted in, for example, donor-funded development projects. We suggest, however, that formal participatory mechanisms are not necessary for technological development. The opportune timing and placement of an innovation can unlock development benefits in a way that any number of formal participative exercises cannot. In cases where markets and choice exist, exit in Hirschman's terms, provides an effective way to determine the appeal of technologies and can bring positive impacts. For example, there were very few formal participation exercises around the introduction of mobile phones in developing countries, and little suggestion that participation exercises were necessary, yet people in the south have adopted this technology rapidly and it is having multiple impacts on the way in which people are able to conduct business and organize life. In these cases where institutional infrastructure allows for choice and competition, the threat of exit (people voting with their purchasing power) is a form of unstructured feedback and may be sufficient to inspire product improvements.

Still, as Hirschman points out, market institutions, in the sense of informal and formal ways of behaving in the buying and selling of goods and services, in both north and south, may be undeveloped or require establishing ways of organizing to facilitate improvement. This is often the case in agricultural and environmental development projects in rural, southern settings where commercial markets are undeveloped. In these cases, people have less opportunity to exit projects through purchasing power. Market analogies are of limited use because 'the market' for these projects often does not relate to conventional market chains and consumers.

In the project examined in the next section, the dynamics of developing a new agricultural technology—tissue culture bananas in Kenya—was driven by the acquisition of funds from development donors. As it was not driven by customer demand, analysis of the development of this new technology necessarily requires examination of the complex non-market-based interactions between the many actors involved, such as national and international NGOs, government agencies, donors and scientists, and their ability to use combinations of exit, voice, and loyalty. Although the stakeholders included the smallholder farmers who were supposed to benefit, and many partnerships and networks evolved to drive the technology's development, a critical, holistic approach—in other words 'good voice'—appeared lacking.

Tissue culture bananas in Kenya

The debate about the appropriateness of agricultural biotechnology has become increasingly divisive and divided in developing countries, and is thus highly politicized. At one end of the spectrum, international NGOs like Action Aid (2003) argue that agricultural biotechnologies should be rejected in their current form, seeing them as a technical fix geared to rich farmers in the north rather than a solution to poverty and inequality in the south (see Chapter 3 for an analysis of the impact of banana commercialization on local livelihoods in south-western Uganda). At the other end, scientists like Wambugu of Africa Harvest claim: 'In Africa genetically modified (GM) food could almost literally weed out poverty'. Biotechnology thus requires the negotiation of a contested terrain of these (and other) ways of thinking (Crush 1995). The development and promotion of tissue culture (TC) banana technologies in east Africa is a case in point.

TC is a relatively unsophisticated biotechnological technique. It is a form of propagation that involves cultivating tissue culture (the growth of plant or animal tissues and cells separate from the organism (adapted from Wikipedia)) in a starter environment in a laboratory, which produces shoot tips that can be planted (Figure 19.1). The resulting banana plants are said to yield more fruit and are resistant to disease. It does not involve genetic engineering, but forms an important building block towards the ability to do this. The technique therefore has advantages. It can act as an indicator of innovative potential and hasten plant breeding leading to improved varieties. TC does however require specific laboratory conditions and training and therefore significant investment in infrastructure and capacity building.

One of the first efforts to promote the idea of tissue culture bananas came from two Kenyan scientists, Wambugu and Kiome (2001). Their assessment of the potential benefits of the technology emphasized the importance of bananas in Kenyan agriculture and in particular the centrality of this crop to smallholder farmers (see also Qaim 1999). In essence this work sought to frame a project that would promote TC banana technology.

Smith (2007), however, has evaluated the studies by Wambugu/Kiome and Qaim critically. Drawing on Latour's (1996) analysis that a project may take on a life of its own and lead to a technology 'push' which may be detached from 'reality', he documents the following sequence in the promotion of the TC banana project in Kenya:

Figure 19.1 Tissue culture banana plantlets in the laboratory.

1. Wambugu and Kiome, and Qaim significantly overestimate the importance of bananas in East African agriculture and in the rural diet in Kenya to the extent that they argue that the TC banana will alleviate poverty and hunger. While cultivating bananas does have advantages, most people grow only a few plants, they are rarely the primary crop, and historically bananas have provided only 11–12 calories to the daily per capita diet over the last 25 years (Food and Agriculture Organisation in Smith 2007).

2. Having established the banana as an important crop in East Africa (albeit on false premises), Wambugu and Kiome document a very serious decline in yield using 'traditional' methods of cultivation, thereby creating what Smith calls a 'crisis narrative'. By this he means their attempt to paint a picture of 'a situation that is inexorable, inevitable and above all cannot be managed with the existing portfolio of development interventions' (Smith 2007: 222).

3. On the basis of the above analysis, advocates of tissue culture bananas raise money and technical input from a wide range of international donors.

4. It appears likely that, in part due to agricultural intensification (increased time spent on orchard management, irrigation etc.), TC bananas do produce increased yield. However, this is not quantifiable in the context of small-scale production and market access.

5. But the TC banana project's success does not rest entirely on its technology. Regardless of its technological merits, it has gained donor support and hence funding.

Figure 19.2 A Kenyan farmer with crop of tissue culture bananas.

The project has also obtained backing from a range of other actors, including the particularly successful and innovative recruitment of small-scale farmers. This is an example of Latour's argument that project success is not inherently attributable to its technical design. Rather, it is a consequence of a project's 'ability to continue recruiting support and so impose . . . [its] growing coherence on those who argue about it or oppose it' (Latour 1996: 78, cited in Mosse 2005).

6. The continuing recruitment of support is sustained despite a later study (Food and Agriculture Organization Data 2004) which shows that both broad data on banana yields in Kenya and data gained from speaking to many small-scale farmers in the areas surrounding Nairobi do not in any way support the disastrous declines in yield originally documented by Wambugu and Kiome.

One of the interesting things about the tissue culture banana work in Kenya is that communication between providers and users of technology was structured into the projects. Organizations such as the African Agricultural Technology Fund (AATF) and the Kenyan Agricultural Research Institute (KARI) are designed to provide important links between technologists and farmers—they are designed to link supply and demand, production, and consumption of technology.

Despite this, research found a mixed reaction amongst farmers involved in the projects and documented considerable disappointment from many of them (cf. Smith 2007). Many farmers adopted the tissue culture bananas in the wake of a collapse in the market for coffee (Figure 19.2). However, even though TC bananas were judged to be of higher quality and to have delivered a durable product, problems with growth cycles of the crop and markets meant that farmers had gluts of bananas that they were unable to sell or consume (Smith 2007; Harsh and Smith 2007). Small farmers interviewed for this research observed 'TC bananas are not meant for local cultivation', 'Kenya needs some mechanism to add value to its bananas', 'No-one thought ahead about surplus bananas' (Harsh and Smith 2007).

Lack of private sector engagement in developing regional or export markets therefore seemed to be a problem in the Kenyan context. Finally, farmers who planted TC bananas were encouraged to increase their investment in banana growing greatly: this involved not only purchasing the TC plantlets (about US$1.50–2.00 per plantlet) but also introducing more labour and capital-intensive orchard management practices such as weeding, irrigation, and intercropping. Without viable markets this clearly left farmers in a vulnerable position.

Towards systemic approaches to participation as institutional development: providing voice to multiple stakeholders

Two points are important in this example. The first concerns the problems inherent in communicating technological benefits in order to drive development 'solutions', whereas a more integrated systemic approach would consider farmers' options and opportunities in a more holistic way. Had initial scoping studies gone beyond promoting research and development into TC bananas as a technical solution, and incorporated a more rounded look at production and marketing constraints, problems may have been identified earlier. In an effort to build on a promising technology and raise money, proponents of TC bananas seemingly both overestimated their importance in terms of subsistence and diet, and ignored some of the production and marketing constraints. Ideally, voice from a range of actors, including the farmers, should have raised these issues, but it didn't, or if it did, it was not heard properly as the project developed its own life.

Thus, the nature of participation in this case has been deficient from both entitlement (the project took on a life of its own where dissenting voices could not be heard) and productivity (its outcomes were deficient) perspectives. Had these been addressed some of the problems might have been avoided. It is important to note that the technology push approach described in the process which actually occurred is very common in developing and distributing technology for development, and particularly so in efforts related to agricultural technologies.

The second point is that the TC banana project work, while flawed, still represents a huge step forward in terms of institutional development. The communications and technology support infrastructure that was set up around these TC banana projects, the involvement of farmer groups and unions and policy people may yet mean that the projects do lead to positive outcomes for farmers. In the wake of initial difficulties this communications work may lead to the development of a physical marketing infrastructure which involves the local private sector and provides farmers with processing and marketing opportunities (Interview with Margaret Karembu, International Service for the Acquisition of Agribiotech Applications, AfriCenter 2004), with export potential. As Hall (2005) points out, innovation projects have a multitude of different starting points and work according to a wide variety of institutional cultures and organizational requirements, none of them ever ideal. The point is to understand what is needed to contribute to their useful evolution. Understanding how participatory approaches and associated communications strategies that allow voice for multiple stakeholders can contribute to the useful construction and evolution of projects and initiatives seems crucial.

■ SUMMARY

- This chapter attempts to think about participation not as an event, but rather as part of an effort to build systems linking S&T innovation with agricultural initiatives in developing countries.

- Two notions of participation are used: entitlement, stressing democratic processes; and productivity, emphasizing achievement of development goals.

- Hirschman's ideas concerning 'exit', 'voice', and 'loyalty' are employed as frameworks to help situate participation in broader institutional contexts.

- The tissue culture bananas case study emphasizes the deficiencies of pushing technological development without adequate critical voice from a range of stakeholders.

- The chapter concludes that the complexity of agents and networks within innovation systems must be examined more holistically, and participation conceptualized within this context, in order to bring about more successful development outcomes.

■ REFERENCES

Action Aid (2003) *GM Crops Going Against the Grain*. London: Action Aid.

Chambers, R. (1997) *Whose Reality Counts? Putting the First Last*. London: Intermediate Technology Publications.

Chataway, J. (2005) Introduction: is it possible to create pro-poor agriculture-related biotechnology? *Journal of International Development* 17(5): 597–610.

Chataway, J. and Smith, J. (2006) The international AIDS vaccine initiative: is it getting new science and technology to the world's neglected majority? *World Development* 43(1): 16–30.

Crush, J. (1995) *The Power of Development*. London: Routledge.

Food and Agriculture Organization (FAO) (2004) *Database for banana yields in Kenya, Tanzania and Uganda. FAOSTAT Dataset 2003*. Available from http://faostat.fao.org

Hall, A. (2005) Capacity development for agricultural biotechnology in developing countries: an innovation systems view of what it is and how to develop it. *Journal of International Development* 17(5): 611–630.

Harsh, M. and Smith, J. (2007) Technology, governance and place: situating biotechnology in Kenya. *Science and Public Policy* 34(4): 251–260.

Hirschman, A. (1970) *Exit, Voice and Loyalty: Responses to Declines in Firms, Organizations and States*. Cambridge, MA: Harvard University Press.

Hirschman, A. O. (1958) *The Strategy of Economic Development*. New Haven, CT: Yale University Press.

Irwin, A. and Wynne, B. (2004) *Misunderstanding Science? The Public Reconstruction of Science and Technology*. Cambridge, MA: Cambridge University Press.

Jasanoff, S. (2007) *Designs on Nature: Science and Democracy in Europe and the United States*. Princeton, NJ: Princeton University Press.

Latour, B. (1996) *Aramis, or the Love of Technology*. Cambridge, MA: Harvard University Press.

Leach, M., Scoones, I. and Wynne, B. (2005) *Science and Citizens: Globalisation and the Challenge of Engagement*. London: Zed Books.

Lundvall, B. (1992) *National Systems of Innovation: Towards a Theory of Innovation and Interactive Learning*. London: Pinter.

Mosse, D. (2005) *Cultivating Development: An Ethnography of Aid Policy and Practice*. London: Pluto Press.

Qaim, M. (1999) *Assessing the Impact of Banana Biotechnology in Kenya*. Brief no. 10. New York: ISAAA.

Smith, J. (2007) Culturing development: bananas, petri dishes and 'mad science'. *Journal of Eastern African Studies* 1(2): 212–233.

Tait, J. (2004) *Science and bias*. Paper presented at the BA Festival of Science, Exeter, UK, 6 September.

Taverne, D. (2005) *The March of Unreason: Science, Democracy and the New Fundamentalism*. Oxford: Oxford University Press.

Wambugu, F. and Kiome, R. (2001) *The Benefits of Biotechnology for Small-Scale Producers in Kenya*. Brief no. 22. New York: ISAAA.

Wikipedia. http://en.wikipedia.org/wiki/Main_Page, accessed 26 January 2009.

FURTHER READING

Hirschman, A. (1970) *Exit, Voice and Loyalty: Responses to Declines in Firms, Organizations and States*. Cambridge, MA: Harvard University Press.

Ruttan, V. (2001) *Technology, Growth, and Development: An Induced Innovation Perspective*. Oxford: Oxford University Press.

Sismondo, S. (2004) *An Introduction to Science and Technology Studies*. Oxford: Blackwell.

Smith, J. (2009) *Technology for Development*. London: Zed Books.

USEFUL WEBSITES

ESRC Innogen Centre. Innogen researches innovation in the life sciences, especially its economic and governance aspects, in both north and south: **http://www.genomicsnetwork.ac.uk/innogen/**.

Open University Centre for Innovation Knowledge and Development (IKD). IKD studies 'dynamics between technology creation and diffusion, business behaviour, government and non-governmental actors': **http://www.open.ac.uk/ikd/index.shtml**.

Open University Development Policy and Practice (DPP) Group. DPP is the development studies group at the Open University, focusing on interdisciplinary innovation in the global context, social learning, and African development: **http://dpp.open.ac.uk/**.

Social, Technological and Environmental Pathways to Sustainability (STEPS). STEPS is a 'global research and policy engagement centre, funded by the ESRC, bringing together development studies with science and technology studies': **http://www.steps-centre.org/**.

20

Designing for sustainability

Robin Roy

Introduction

Designing for sustainability immediately raises two questions: what is designing and what is sustainability?

The creation of every new object or system involves designing. Designing means creating the ideas, concepts, sketches, drawings, computer models, prototypes, and instructions to enable something new or improved to be made—such as a dress, chair, car, or building. Designing has taken place for thousands of years but it is only since the Industrial Revolution that the modern design professions—engineering design, industrial design, graphic design, interior design, exhibition design, and so on—have emerged. Before the Industrial Revolution the designer was often also the maker; a craftsperson who copied or adapted traditional designs—such as a plough or farm cart (Sturt 1993). Since then designers, engineers, and architects have become the creative intermediaries between the consumer or client and the process of manufacturing or construction.

Sustainability is often viewed as having social, economic, and environmental dimensions. Environmental sustainability requires that the production and consumption of goods and services does not exceed the earth's capacity to support existing populations and that of their children, grandchildren, and so on, into the indefinite future. It has been estimated that to allow the growing populations of developing countries to reach decent living standards, while moving towards sustainability, will require energy and resource consumption and emissions generated per person in industrialized countries to fall by up to 90 per cent from present levels during this century (United Nations Environment Programme 1999: 12).

Although there are many urgent environmental issues, a recent focus has been on reducing emissions of climate-changing gases arising from energy use and the production and consumption of food, goods, and services. For example, the authoritative UK Government Stern Review (HM Treasury 2006), said that the scientific evidence indicate a 50 per cent global reduction, and a 60–80 per cent reduction in developed countries' greenhouse gas emissions is needed by 2050 to reduce the risk of the world suffering dangerous climate change. Subsequent reports indicate that these reductions are the minimum required (e.g. Committee on

Climate Change 2008). At the same time concerns have re-emerged about how long economic supplies of oil, gas, and water will last given population and consumption growth.

Designing for environmental sustainability

How can such large and rapid reductions in resource consumption and emissions be achieved? One way is to consume less; but that is not something that the citizens of industrialized—let alone the growing middle-class of developing countries—are so far willing to accept. The other way is to produce and consume goods and services that have less impact on the environment; in other words to design for environmental sustainability (DfS). DfS means creating concepts and configurations of materials, components, and technologies for new or improved products, buildings, and systems that minimize energy and materials use, emissions, and waste. To be acceptable to consumers these items must also meet the usual design requirements of good performance, attractive appearance, safety, reasonable cost, ease of production, etc.

It is possible to see DfS as very old or relatively new. Historically, and in parts of developing countries today, people designed and made their artefacts and buildings from local, natural, or renewable materials such as clay, straw, and wood. Or if they used valuable non-renewable materials such as gold or jade the objects were designed to last for hundreds of years. Using natural and renewable materials and/or creating long-lasting products are also contemporary DfS approaches, for example building houses using straw bales or moulding a chair from crushed almond shells.

The British Arts and Crafts movement of 1850 to 1914, led by individuals like William Morris, criticized ugly factory-made goods and the pollution and human exploitation of the Industrial Revolution, advocating a revival of handicrafts and a return to a simpler, rural way of life in harmony with the environment. In contrast, the Modernist movement in design and architecture in the 1920s embraced urban living, industry, and technology. It adopted many utopian socialist ideas of the Arts and Crafts movement, but emphasized machine-like efficiency, strict functionality, and economy of materials in design. Although Modernists were not explicitly environmentalists, the design philosophy behind Marcel Breuer's lightweight tubular steel furniture and Mies van der Rohe's motto 'less is more' fits well with modern DfS principles (Faud-Luke 2005).

Contemporary DfS really began with the Alternative Technology (AT) movement that sprang from the environmental concerns and the counter culture of the 1960s and 1970s. Alternative technologists proposed radical designs and innovations such as self-sufficient homes and communities independent of central sources of energy, water, and food. A specific branch of AT was 'appropriate technology', which aimed to generate employment in developing countries, while being low in capital cost, easily produced and maintained, and environmentally safe (Carr 1985). As in the earlier design movements, the innovations of Alternative and Appropriate Technology were linked to social and economic change; and hence also labelled 'design for the real world' (Papanek 1985) and 'radical technology' (Harper and Boyle 1976). In other words, DfS has its roots in the view of sustainability as having environmental, social, and economic dimensions.

DfS is therefore not new. However, it has evolved from something people did because of materials and technology they had available, or as part of a critical social movement, to a mainstream activity advocated by governments, required by legislation and practised by industry (Tukker *et al.* 2001; DEFRA 2008). How did this come about? The initial response of industry to environmental problems of the 1970s was to install 'end-of-pipe' pollution control equipment, for example to clean waste water or filter emissions from car exhausts. The next step was to install systems to allow products to be manufactured with less energy and waste. Eventually some companies realized that environmental impacts mainly depended on the *design* of products or systems (their materials, energy use, recyclability, etc.) and by the 1980s and early 1990s, the idea of designing 'greener' products emerged.

Since then the pressures on designers, architects, and industry to design for the environment have increased; driven in Europe by a growing body of legislation. For example, the European Union's (EU) energy label, introduced in 1996, requires that the energy efficiency of appliances such as refrigerators is rated and displayed on a label. Following a 2006 EU Directive, all homes sold or rented must be rated for their energy efficiency and carbon dioxide (CO_2) emissions and given an Energy Performance Certificate. Another important piece of EU legislation was the 2006–07 Waste Electrical and Electronic Equipment (WEEE) Directive that aims to reduce electrical and electronic waste by requiring manufacturers to design products for recycling and reuse and inform consumers how to recycle waste electrical goods. A further factor encouraging manufacturers to design for the environment is the concern that competing firms will gain an advantage by designing greener products. Hoover, for example, designed its first range of energy- and water-saving washing machines in response to German competition and consumer demand (Roy 1997). Finally, pressures from green investors and from **corporate social responsibility** (CSR) policies can be important drivers. Almost all large businesses nowadays have an environmental policy, whether mainly for public relations or for genuine social responsibility reasons.

DfS—strategic approaches

So far DfS has been discussed as if it was one thing. But there are many different approaches, which have been categorized into four levels (Brezet 1997)—labelled 'green design', 'eco-design', 'sustainable design' and 'sustainable innovation' (Roy 2006). The first two approaches focus on environmentally sustainable design, while the second pair also include social and economic sustainability. They are considered in turn below.

Green design

Green design means designing products that have a reduced load on the environment, but with a focus on one or two issues. In practice, most green design projects focus on one or more of the following three environmental issues:

• Materials conservation: for example, furniture or clothing made from materials that would otherwise be wasted (Figure 20.1a); pencils made from recycled plastic cups.

Figure 20.1 Green design: (a) clothing items made from waste can ring pulls; (b) 'Dr Skud' fly swat, designed by Philippe Starck.

- Energy conservation: for example the compact fluorescent lamp (CFL), that uses about a quarter of the electricity of an incandescent light bulb and also lasts about eight times as long, so conserving materials as well as energy.
- Toxic or hazardous emissions avoidance: for example, an aerosol that uses compressed air instead of inflammable hydrocarbon gas propellants; the 'Dr Skud' fly swat created by the famous French designer Philippe Starck as a desirable and fun product that eliminates the need for chemical flysprays (Figure 20.1b).

Eco-design

Green design accounts for most DfS projects and is the focus of much legislation, such as the WEEE Directive (see above), but it has a major drawback; an environmental gain in one area can produce a loss in another. For example, if a product is made from recycled materials it may be heavier, needing more energy to transport it, or be less durable and hence quickly discarded. Eco-design attempts to deal with that drawback. Eco-design is often called 'life-cycle design' because it is based on a technique called **life-cycle assessment** or analysis (LCA), which attempts to assess a product's 'cradle to grave' environmental impacts from the extraction of its raw materials, through production, distribution, and use, to its disposal or recycling. If a design team knows the impacts at the different life cycle stages it can try to reduce the largest ones and balance one impact against another.

For example, early attempts at designing a 'greener' electric kettle typically focused on facilitating disassembly and recycling at the end of the kettle's life. However, a LCA clearly revealed that a kettle's greatest impacts, notably greenhouse gas (GHG) emissions and power station waste from generating electricity, occur during the *use* phase of its life cycle. This knowledge stimulated a design team to eco-redesign a kettle. They reduced their eco-kettle's energy use and GHG emissions by 25 per cent relative to a previous model by insulating the kettle walls and placing indicators on the lid to show how much water was needed for a cup and whether the water remained hot enough to make tea or coffee. In addition, the team took the opportunity to reduce the material content and increase the proportion of recyclable plastics in the eco-kettle (Sweatman and Gertsakis 1996).

Life cycle eco-design is a powerful DfS approach, but also has drawbacks. First, LCA is time-consuming and expensive. So in practice, most designers use a simplified eco-design approach based on life-cycle thinking. For example, the multinational electronics firm, Philips, tried to introduce life cycle eco-design throughout its business, but found it too complex. Instead Philips introduced a simple checklist based on five 'green focal areas'— reducing weight; eliminating hazardous substances; reducing energy consumption; more recycling; and less packaging—that help its designers and engineers to consider the important impacts of a product. For example, using green focal areas Philips' lighting designers reduced the size, weight, and toxic mercury content, improved the efficiency and established recycling systems for its range of fluorescent tubes (Philips 2005).

Sustainable design

A more important deficiency of both green design and eco-design is that neither is going to be enough to achieve the necessary reductions in industrialized countries' resource use and emissions to move towards global environmental sustainability. Green and eco-design can help, but even if everyone used pencils made from recycled plastic cups, sat on chairs made from waste wood, and used eco-kettles, consumer societies would still be living unsustainably. Over two-thirds of resource use and emissions in industrialized countries arise from household consumption, including housing, personal travel, and the supply of food, goods, and services. The rest of the chapter focuses on UK housing and in particular its energy use and CO_2 emissions.

In Britain, homes use nearly a third of delivered energy and produce 27 per cent of total CO_2 emissions (CLG 2006a). Sustainable design for households includes improving the environmental performance of existing homes and building more sustainable new housing. To make Britain's 25 million existing homes more sustainable means making improvements like insulating lofts and walls, installing double- and triple-glazed windows, collecting rainwater, avoiding timber from non-sustainable forests, etc. As important as upgrading the building itself is improving the environmental performance of its heating, lighting, and appliances. This means replacing existing heating systems with energy-efficient condensing central heating boilers, or more innovative systems such as heat pumps (which extract heat from soil or water), and wood pellet stoves; using energy-efficient appliances and new lighting technologies, such as light-emitting diode (LED) lamps that seem likely to displace CFLs in the future.

Systems to generate energy at the household level can also be installed in existing homes, including roof-mounted photovoltaic (PV) cells that generate electricity from the Sun's energy and micro combined heat and power units that generate both electricity and heat from a variety of fuels. At present such 'microgeneration' systems are quite rare in the UK because of their cost, novelty and complexity, but are likely to become much more widespread in the future (Roy *et al.* 2008). Given favourable government support, there could be 9 million such systems in British households by 2030 (Element Energy 2008).

Although there are many improvements that can improve the sustainability of existing housing, there are limits. Not everyone can afford to make the required changes without subsidies. Also there are forces acting against sustainability, including the growth in single occupancy households that use more resources per head than multiple person households, consumers buying bigger homes and larger appliances, and new energy-hungry items like digital TV and heated conservatories. It has been estimated that making socially and economically feasible improvements to existing UK housing will reduce energy use and CO_2 emissions by 25–35 per cent by 2050, well short of the 80–90 per cent cut required for long-term sustainability.

As well as the potential for upgrading existing housing, Britain constructs about 200,000 new homes per year. Most are additional rather than replacements for existing housing, only a few thousand of which are demolished each year (Boardman *et al.* 2005). This new housing is more sustainable because of tighter building regulations, but most new UK homes are far from what can be achieved, as is demonstrated by examples of sustainable housing design. One pioneering example of sustainable housing is the Autonomous Urban House built in 1993 near Nottingham, England. It is almost entirely solar heated, independent of mains water and sewerage, while photovoltaic (PV) cells provide about half its electricity. Total energy use is about 10 per cent of a typical UK house. However, its cost was fairly high and living in it requires some commitment, such as opening and closing windows to maximize solar gain, and emptying the composting toilet (Vale and Vale 2000). It is possible therefore to design sustainable housing that achieves the 90 per cent reductions in resource use and emissions, but such homes are rare and most developers and house buyers would not want to attempt the sustainability levels of the Autonomous Urban House.

Sustainable innovation

One way of achieving sustainability for a greater number of people in industrialized countries is through sustainable innovation. This means going beyond designing individual products or buildings to developing whole systems that are environmentally, socially, and economically sustainable. A pioneering example of this approach is the Hockerton housing project, also built near Nottingham, that takes the autonomous house concept further to the small community level, thus sharing some of the cost and effort of building and living sustainably. The Hockerton project's five family homes, completed in 1998, look distinctive because of the solar conservatories across the terrace front and earth covering over the rear. The interiors are light and modern and stay warm all year with no artificial heating because of the passive solar design and super-insulation. Total energy use for each unit was monitored at 25 per cent of an average UK home (BRE 2000). Subsequently, two wind turbines and PV cells were installed for generating electricity, with the result that each household's energy demand is now only some 10 per cent of the UK average. In addition, Hockerton residents collect all their own water, process sewage in a reed bed (see below) and grow some of their food. Each household is allowed one conventional car, which they try to share, while cycles and an electric car provide local personal transport. Some of the residents earn their living on-site while others have conventional jobs. In other words, the project aims to be socially and economically as well as environmentally sustainable.

Reed bed systems provide a small-scale and environmentally unobtrusive form of sewage treatment. The bed is usually a rectangular area filled with gravel and planted with the common reed, *Phragmites australis*. Sewage is trickled across and through the bed where the roots of the plants allow contact between the waste water and oxygen in the air. This creates conditions conducive for the bacteria that feed on the organic matter in the sewage and reduce its polluting potential. The resulting effluent should be clean enough to discharge to rivers without causing harm.

An example of sustainable innovation at a bigger scale is the BedZED (Beddington Zero Energy Development) completed in 2002 in South London. BedZED houses about 200 people in 82 houses and flats (Figure 20.2). Where possible local or recycled construction materials were used. The heat-recovery ventilation system plus the solar conservatories and heavy insulation mean heating BedZED units uses only about 12 per cent of the UK energy average. Waste water and sewage is treated via a greenhouse reed bed and used in toilets and gardens; so mains water demand is about half the UK average. An attempt has also been made to reduce car use by locating the development near public transport, restricting parking, and having a car-sharing club. Although the main rationale for BedZED is very low CO_2 emissions, it also aims to be economically and socially sustainable with its onsite work spaces, social housing units, nursery, and sports field.

Figure 20.2 BedZED sustainable housing, South London. BedZED's distinctive features include solar conservatories, first-floor gardens, and brightly coloured roof ventilators that draw in fresh air and expel stale air via a heat exchanger.

BedZED has been successful; its light-filled interiors, conservatories, small gardens, energy- and water-saving features are popular. However, each home is relatively small and construction costs were higher than for conventional housing. Also the combined heat and power (CHP) plant fuelled by waste timber and the rainwater collection system have encountered technical problems. Sustainable housing projects like BedZED show what can be achieved, while its higher costs and technical bugs are not surprising given the degree of innovation it involved.

Since BedZED was completed there has been an increased interest in sustainable housing, stimulated by the 2007 UK government announcement that by 2016 all new homes should be 'zero-carbon', meaning that over a year each dwelling or development should produce net zero CO_2 emissions. The strategy for achieving zero-carbon is the Code for Sustainable Homes (CfSH), which sets CO_2 reduction targets, and awards points for other features including water saving, low-impact building materials, and waste recycling (CLG 2006b). The Code has resulted in major UK house-builders designing prototype zero-carbon homes and housing developments, while BedZED's architects have designed a timber frame kit house with heat-retaining concrete walls and ceilings that can be specified to meet the various levels of the CfSH up to the zero-carbon standard (Dunster *et al.* 2008).

There are also radical proposals applying ideas pioneered in BedZED in larger sustainable community projects. For example, an eco-city at Dongtan near Shanghai, China is being designed, although on a smaller scale than originally planned (see Chapter 21), while sustainable communities in several countries are planned as part of a programme called One Planet Living based on the idea that each country's population has to live sustainably within the earth's resources (Bioregional 2008).

■ SUMMARY

- Designing for sustainability (DfS) can range from designing a pencil made from recycled plastic cups to a zero emissions, zero waste community for thousands of people.

- All DfS levels are necessary, but to reach the 80–90 per cent reductions in energy and resource use and in emissions and waste up to 2050 and beyond, this chapter has argued it is necessary to design at the sustainable innovations level and that designing individual greener products will not be enough.

■ REFERENCES

Bioregional (2008) *One Planet Living: Communities*. Available from http://www.bioregional.com/programme_projects/opl_prog/communities.htm.

Boardman, B., Darby, S., Gillip, G., *et al.* (2005) *40% House*. Environmental Change Institute, Oxford, February. Available from http://www.eci.ox.ac.uk/research/energy/40house.php, accessed August 2008.

BRE (2000) *The Hockerton Housing Project. Lessons for Developers and Clients. New Practice Profile 119*. Watford: Building Research Establishment.

Brezet, H. (1997) Dynamics in ecodesign innovation, UNEP. *Industry and Environment* 20(1–2): 21–24.

Carr, M. (ed.) (1985) *The AT Reader. Theory and Practice in Appropriate Technology.* London: Intermediate Technology Publications.

CLG (2006a) *Review of Sustainability of Existing Buildings. The Energy Efficiency of Dwellings—Initial Analysis.* London: Department for Communities and Local Government, November.

CLG (2006b) *Code for Sustainable Homes. A Step-change in Sustainable Home Building Practice.* London: Department of Communities and Local Government, December.

Committee on Climate Change (2008) *Building a Low-carbon Economy—The UK's Contribution to Tackling Climate Change.* London: The Stationery Office.

DEFRA (2008) *Progress Report on Sustainable Products and Materials.* London: Department of Environment, Food and Rural Affairs. Available from http://www.defra.gov.uk/environment/consumerprod/, accessed August 2008.

Dunster, B., Simmons, C. and Gilbert, B. (2008) *The ZED Book. Solutions for a Shrinking World.* Abingdon: Taylor and Francis.

Element Energy (2008) *The Growth Potential for Microgeneration in England, Wales and Scotland. Report Commissioned by BERR and Others.* Available from http://www.berr.gov.uk/energy/sources/sustainable/microgeneration/research/page38208.html, accessed August 2008.

Faud-Luke, A. (2005) *The Eco-design Handbook,* 2nd edn. London: Thames and Hudson.

Harper, P. and Boyle, G. (eds) (1976) *Radical Technology.* London: Wildwood House.

HM Treasury (2006) *Stern Review: The Economics of Climate Change.* London: HM Treasury. Available from http://www.hm-treasury.gov.uk/sternreview_index.htm, accessed August 2008.

Papanek, V. (1985) *Design for the Real World,* 2nd edn. London: Thames and Hudson.

Philips (2005) *Sustainability Report 2004.* Royal Philips Electronics, available from http://citybeautification.com/gl_en/environment/pdf/sustainability_report_2004-13939.pdf, accessed August 2008.

Roy, R. (1997) Design for environment in practice—development of the Hoover New Wave washing machine range. *Journal of Sustainable Product Design* 1(April): 36–43.

Roy, R. (2006) Products: New Product Development and Sustainable Design. *T307 Innovation: Designing for a Sustainable Future, Block 3.* Milton Keynes: The Open University.

Roy, R., Caird, S. and Abelman, J. (2008) *YIMBY Generation. Yes in my back yard! UK Householders Pioneering Microgeneration Heat.* London: Energy Saving Trust. Available from http://design.open.ac.uk/research/research_dig.htm, accessed August 2008.

Sturt, G. (1993) *The Wheelwright's Shop.* Cambridge: Cambridge University Press.

Sweatman, A. and Gertsakis, J. (1996) Eco-kettle: keep the kettle boiling. *Co-Design* 05-06: 97–99.

Tukker, A., Charter, M., Haag, E., Vercalsteren, A. and Wiedmann, T. (2001) Eco-design state of implementation in Europe. *Journal of Sustainable Product Design* 1: 147–161.

United Nations Environment Programme (1999) *Global Environmental Outlook 2000.* London: Earthscan.

Vale, B. and Vale, R. (2000) *The New Autonomous House: Design and Planning for Sustainability.* London: Thames and Hudson.

■ **FURTHER READING**

Brezet, H. and van Hemel, C. (1997) Ecodesign: a promising approach to sustainable production and consumption. *UNEP Industry and Environment* 20 (1 and 2).

Lewis, H., Gertsakis, J., Grant, T., Morelli, N. and Sweatman, A. (2001) *Design + Environment, A Global Guide to Designing Greener Goods.* Sheffield, Greenleaf Publishing.

Datschefski, E. (2001) *The Total Beauty of Sustainable Products*. Hove: Rotovision.

Fuad-Luke, A. (2005) *The Eco-design Handbook,* 2nd edn. London: Thames and Hudson.

■ USEFUL WEBSITES

BioThinking, developed by green entrepreneur Edwin Datschefski, means looking at the world as a single system, and developing new ecology-derived techniques for industrial, organizational, and sustainable design: **http://www.biothinking.com/**.

BioRegional is a charity which invents and delivers practical solutions for sustainability, e.g. The One Planet Living Programme of sustainable communities: **http://www.bioregional.com/**.

Demi is a web-resource, developed at Goldsmiths College, University of London, bringing together wide-ranging information on design for sustainability: **http://www.demi.org.uk/**.

An ecodesign resource from Loughborough University developed to support designers who want develop more environmentally and socially responsible products: **http://www.informationinspiration. org.uk/**.

Brings together UK government work on product life-cycle assessment, product information, and evidence on sustainable consumption and production and waste: **http://www.defra.gov.uk/ environment/consumerprod/**.

An international network of designers and others for anybody interested in sustainable design: **http://www.o2.org/index.php**.

SusProNet was a network of industries and institutes supported under the EU Fifth Framework Programme. SusProNet developed and exchanged expertise on design of product-service systems for sustainable competitive growth: **http://www.suspronet.org/**.

The Centre for Sustainable Design facilitates discussion and research on eco-design and sustainability in product and service development. The Centre also acts as an information clearing house and a focus for innovative thinking on sustainable products and services: **http://www.cfsd.org.uk/**.

SusHouse was a European research project concerned with developing and evaluating scenarios for transitions to sustainable households: **http://www.tbm.tudelft.nl/live/pagina.jsp?id=dfe511c2– 82ea-47d2-a6fa-829fc60950a3andlang=en**.

21

Corporate responsibility in practice: building an eco-city from scratch

James Warren

Introduction

This chapter is based on one company's pioneering work in developing resource management systems and integrated design thinking to make cities more sustainable. The case of Dongtan eco-city near Shanghai in China is used as an example to illustrate the types of frameworks and guiding principles which have been used to design an eco-city. The company is Arup, named after its founder Sir Ove Arup. It is a global firm of designers, engineers, planners, and business consultants providing a diverse range of professional services to clients around the world.

Dongtan was chosen by the Shanghai Industrial Investment Corporation (SIIC) to be the site for an impressive demonstrator eco-city and in mid 2005 Arup was contracted to design and masterplan the project. As an eco-city designed and built from the ground up, Arup had to take an innovative approach to formulating the key principles for guiding the building phase of the project. The site chosen covers about 8,400 hectares on the south-east tip of the island of Chongming which lies in the mouth of the Yangtze River a few hours by bus and boat from Shanghai (Figure 21.1). It is an agricultural area adjacent to a wetland of ecological importance that provides a stopping point for about one million birds every year on the major bird migration route known as the East Asian–Australasian Flyway (Normile 2008). Recently completed transport links to the mainland have made Dongtan more accessible with faster travel times and may make it appealing as a tourist destination.

What is an eco-city? Generally one might say an eco-city is a city which is more sustainable than a traditional one, or a city that uses far fewer resources when compared to a traditional city. Often sustainability is seen principally as an environmental concept, but

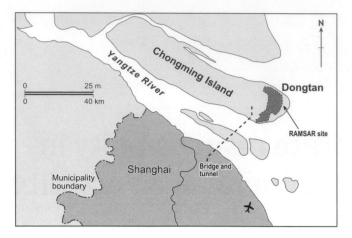

Figure 21.1 Map of China's east coast showing location of Dongtan site off the coast of Shanghai. Adapted from Wood (2007).

a true eco-city must be sustainable environmentally, socially, economically, and culturally (Arup 2008a). As Green (2008) has written of the Dongtan project:

Dongtan is being designed to be self-sufficient in energy, food and water; efficient in waste management, and responsible—potentially—for zero carbon emissions from transport. If the promise of Dongtan is realised, it would demonstrate the clear benefits of limiting pressures on nature, increasing resource productivity, and aligning the needs of people with those of the environment.

Design of an eco-city is therefore a complex task that needs to consider all these factors and the interactions between them. This chapter attempts to summarize some of the processes and approaches for incorporating sustainability that were used in the Dongtan project and have been developed as generic tools for the design of other potential eco-cities.

Design processes

In order to meet the sustainability goals, Dongtan's many project designers had to grapple with multiple demands including engineering targets and resource constraints for land, water, energy, waste, and transport. These constraints and targets had to be kept in mind whilst simultaneously aiming to achieve environmental, social, and economic development within an overarching sustainability remit. Not surprisingly any improvement towards meeting one objective needed to be considered against all others to ensure there were the least number of detriments (Wood 2007). The design of a holistic sustainable eco-city emphasizes not only the importance of considering the fundamentals, such as finite resources, but also the critical nature of the social fabric and the importance of culture and cohesion within cities. All of these considerations needed to be incorporated in the design process.

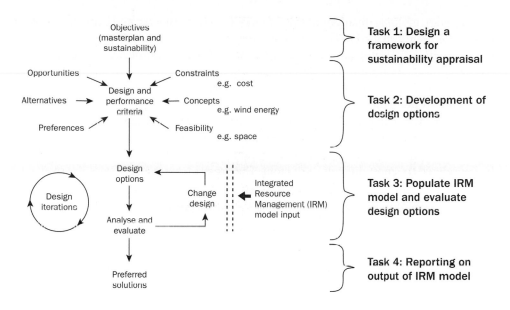

Figure 21.2 A diagram showing the main tasks required in undertaking the Dongtan eco-city development. Adapted from Wood (2007).

An outline of the process used for Dongtan is shown in Figure 21.2, portrayed in a linear fashion with four main tasks, which are described below.

Task 1 was to build the overall masterplan of the eco-city and create a framework for sustainability appraisal. This meant deciding what the main objectives were within the project and how these might begin to be realized. One might think of it as the highest level vision.

For Task 2 a variety of tools were used to develop design options. For Dongtan, this required consideration of all the diverse elements of the project such as cost, wind energy availability, and space, among many others. It was also necessary to identify the inter-relationships between all the different factors, quantify the areas where trade-offs would be necessary and examine the impacts of those trade-offs. Trade-offs, where improvement in one factor was coupled with detriment in another, were sometimes inevitable due to constraints such as time, availability of materials, and cost. It was important that where the trade-offs between competing objectives were unavoidable, they were openly acknowledged and minimized (Wood 2007).

Task 3 was centred on Arup's IRM (Integrated Resource Management) computer model. The IRM model is a systems analysis tool that adopts a **life-cycle assessment** (LCA) approach and applies this to regional or town/city masterplans. The model was designed to assess the sustainability performance of the design options generated in Task 2 and find optimums for a large number of variables. The interconnectivity within this model is critical. Calculating passenger transport energy, for example, is dependent on a broad range of variables such as travel demand, housing density, freight movement, population, and others. Linked variables such as these interact and influence each other and by modelling them all, the IRM model allows the systematic analysis and evaluation of sustainability performance.

For Dongtan, multiple iterations of the IRM model were considered and comparisons with other cities were also used in order to determine 'typical' values as well as best practice values for metrics like water and energy consumption. The model provided the project team with the information they needed to decide which design options were likely to produce the best results so that, following the reporting stage in Task 4, the preferred options could be taken forward to the actual building site.

In practice these four tasks overlap and involve multiple iterations, but the linear flow diagram in Figure 21.2 is useful for defining the approximate boundaries of each major task. If Task 1 seems rather 'short' in the diagram, be assured that in some ways Task 1 can be considered the most difficult of the four. The project team had to devise the overarching principles which were to guide them throughout the entire eco-city building phases, and in theory last until 2040, or 2050, when Dongtan was predicted to grow to 500,000 inhabitants.

The top level vision of Task 1, for Dongtan, resulted in eight key principles which are intended to be revisited throughout the life of the project to act as a guide towards achieving the eco-city ideals. The eight principles of the Dongtan sustainability framework are:

- to preserve the wetland habitat;
- to create an integrated, vibrant and evolving community;
- to improve quality of life and create desirable lifestyles;
- to create an accessible city;
- to ingrain contemporary Chinese culture into the city fabric;
- managing the use of resources in an integrated manner;
- working towards carbon neutrality; and
- utilizing governance to achieve long-term economic, social, and environmental sustainability.

These key principles were derived after many workshops and meetings of the various stakeholders including the client, SIIC, Arup's engineers and designers, demographers, academics, NGOs, and others. Ownership of these principles of sustainable development among all the stakeholders, as well as the process used to achieve them, was considered to be important in making the eco-city work.

The Dongtan network workshops also resulted in 18 key objectives which the stakeholders set as their reference point. The objectives were grouped under four headings: environment, natural resources, societal, and economic. There is not room here to include all 18 objectives but some examples are:

- Physically and legally protect the eco-city's internationally significant wetlands from any man-made intervention such as physical encroachment, poaching of wildlife, and pollution to land water and air; impose the strongest possible penalties on organizations and individuals in breach of these measures.
- Design for the reduction, reuse, and recycling of natural and man-made materials and develop policies that encourage resource management, sustainable production and consumption, and the extraction of the maximum benefit from residual wastes at the eco-city, through energy production and use in agriculture.

- Ensure all citizens can engage with and are represented by governance systems that are accountable and that work towards the continued realization of the fullest concepts of the eco-city.

- Aim for consistent economic progress, which recognizes China's old and new economies and allows for the sustainability objectives of the eco-city to be met.

The 18 objectives were devised in order to guide decision making in the design process. In order to visualize the complex, multifactorial considerations of the process that are embedded within the objectives another computer model was used. The key factors were mapped using a visual modelling tool called SPeAR® (Sustainable Project Appraisal Routine), which works in conjunction with the IRM model. SPeAR® acted as the main information management tool and reporting device throughout the Dongtan design process (ARUP 2008b). It encouraged all participants to consider how each change in the masterplan could affect the various parameters in the city and how the key objectives might best be achieved.

The SPeAR® computer model uses input data for a wide range of variables and has been programmed to calculate how these factors may interact and what the outcomes will be. The outcomes are quantified and rated, from optimum to worst case, according to their impact on sustainability. The different options and choices are then mapped on to a four-quadrant framework using the same four headings adopted for the Dongtan objectives, and the results are displayed as sectors in a target-shaped diagram (see Figure 21.3). The model is bespoke and can be adapted for any project or product that is being assessed. For example, in Figure 21.3 the diagram has 22 sectors but this number may range from 15 to 25 depending on the project. Each sector may be derived from several different variables, for example, in the case of air quality there may be as many as ten, making the total number of measures or key indicators incorporated in the model quite large.

The SPeAR® model is a powerful tool for visualizing and comparing different options. The 'target' gives participants in the process a visible picture of various options in an easily accessible and meaningful way. Optimum design options will result when the greatest number of sectors within the overall circle are in the darkest shade, heading towards the centre of the target. Overall, SPeAR® allows users to highlight strengths and weaknesses. It points to future areas of opportunity, provides management information to aid decision making and provides auditable information for assurance and verification purposes (ARUP 2008b). The entire process is ultimately asking the question 'how green is the city?' under stated conditions.

Design options

The modelling tools can come up with some unexpected surprises. For Dongtan, when the transport system was modelled using fuel cell-powered vehicles that were very quiet, the IRM results indicated that energy consumption in buildings would drop as a consequence. The connection between these two factors, i.e. quiet vehicles and energy use in buildings, is not immediately obvious, but the model predicted that with quiet streets, people would be more likely to open their windows to get fresh air, so would use less

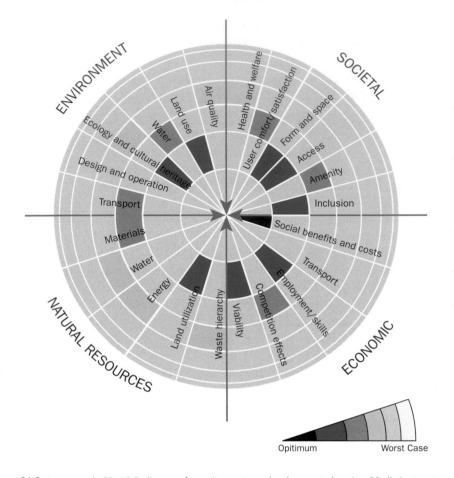

Figure 21.3 An example SPeAR® diagram for a city centre redevelopment showing 22 distinct sectors (Arup 2008b). The sectors to be included will vary for different projects.

air-conditioning and therefore less energy. However, the model did not emphasize the importance of vehicle engine sound in the overall safety of pedestrians and cyclists. In later iterations, the designers began to consider the need to incorporate some noise, at a low level, to warn pedestrians and others of traffic. There are many other examples where the modelling demonstrated the links between factors and showed how changes in one could lead to unexpected consequences in another.

At the time of writing the implementation of Arup's Dongtan masterplan has been indefinitely postponed. There are only artists' visions of what the city might look like, but also quite a few design considerations which suggest that the city will be relatively low-rise with buildings no more than 6–8 stories tall. This will optimize solar gain (to capture heat from the sun) and photovoltaic positioning (to generate electricity) (Arup 2006). The infrastructure will be centred around the inhabitants' perspective and be pedestrian-centric with an emphasis on bicycle and public transport as the 'basic unit' rather than the car. Imagine a city where every single dwelling is within a 7-minute walk away from either a public transport stop or the nearest car pool or car-sharing hub. Shops and amenities are close by, also reducing trip numbers. The buildings will have turf and vegetation roofs to act as natural insulation layers and to reduce run-off for optimum water use. Waste is recycled at very high levels with buildings and neighbourhoods designed with waste collection and reuse in mind, rather than added on as an afterthought. Arup claims that one of the most important aspects of the project is the holistic approach being taken that considers every issue from many perspectives (Normile 2008). By doing this in such a systemic manner it is hoped that the community will be cohesive and appreciative of the work behind many of the plans.

Conclusion

Dongtan acts as a demonstration of the ideal city with the eyes of the world watching the development to see if it will succeed in its ambitions of making the best possible use of energy, water, and materials (Normile 2008). Other model cities are planned, both in China (such as Wanzhuang), and elsewhere (such as Masdar City in Abu Dhabi), and although some of the 'eco-demonstrator' principles may be similar the various options and solutions applied on a city-wide scale will surely be different. Even if Dongtan becomes a global showcase of excellent urban design and technology, these design options may not be affordable or relevant elsewhere in China (Economist 2006). As Peter Head (Director, Arup) says:

> Eco-cities need to be delivered on the ground at community level, as in Dongtan . . . there are no blueprints for a typical eco-city. We need to find city-specific solutions that provide a higher quality of life at lower ecological cost, and help cities deal with risks such as climate change and access to clean water and food.

> (Head 2008a)

This suggests that transplanting ideas from one city to another may not be easy or straight-forward in terms of retrofitting practices, but that is not a valid reason to continue down a pathway of maximum consumption rather than that of optimum consumption (Head 2008b: 41).

This chapter has explored what eco-cities might offer as blueprints for cities of the future. Both Dongtan and other eco-cities offer opportunities for learning and ways of creating more sustainable cities worldwide. In many ways the learning is about sustainability, rather than, as some might see it, simply about masterplans. In fact the Dongtan case study shows that the design and execution of city construction is far more complex than simply the roll-out of the best-laid masterplans. City planning also has to cope with complex change over time including shifting demographics and economics. One could argue that one of the main reasons for building an eco-city is to gather information and learn how to build better cities in the future, and that Dongtan is just a step in a much longer process. In this way the next eco-city projects will benefit greatly from the knowledge learned there. Perhaps most importantly we must recall that sustainability is not static, and although some of the complex models and tools provide an information-rich analysis for specific points in time, these snapshots must be revisited periodically. Only by doing this does our practice in sustainability become an ongoing process which is in itself inherently sustainable.

■ SUMMARY

- Eco-cities have recently captured the attention of many people around the globe as a possible way for city living that has lower impacts.

- Dongtan is one of the planned eco-cities for which the Arup consultancy company are planners and city designers.

- Eco-city 'design, build, and operate' involves a highly complex set of interlinked tasks, which require multiple computer-driven tools and simulations in order to scrutinize all of the various options.

- With lessons learned, eco-cities, whether retrofit or from scratch, may offer huge reductions in energy, waste, water, and food/consumables.

■ ACKNOWLEDGEMENTS

My thanks go to Beth Hurran, Sally Quigg, Neil Kirkpatrick, and Roger Wood of Ove Arup and Partners (London) for their time and support. Thanks also to A. J. Lloyd for cartographic services.

■ REFERENCES

ARUP (2006) *Case study: sustainable city—Dongtan*. Available from http://www.c40cities.org/docs/casestudies/buildings/dongtan_carbon.pdf.

ARUP (2008a) *Dongtan Eco-City Briefing Document*, Arup Press Office. London: Arup.

ARUP (2008b) SPeAR ®: Product Overview. Available from http://www.arup.com/environment/feature.cfm?pageid=1685, accessed 28 October 2008.

Economist, The (2006) Visions of ecopolis. *The Economist* 380(8496); 20–23, Technology Quarterly.

Ove Arup and Partners Ltd (2006) *Sustainability Appraisal for Thornfield Properties plc, Silver Hill Renaissance, April 2006, Final Report, Job Number 113666*. Available from http://www.silverhill-winchester.info/pdf/Environmental_Statement_Appendices_6.pdf, accessed 28 October 2008.

Green, S. K. (2008) *The Dongtan Ecocity Project – the Birth and Co-evolution of Capabilities, Research Project Imperial College Business School*. Available from http://www3.imperial.ac.uk/business-school/research/projects/dongtan, accessed 28 October 2008.

Head, P. (2008a) *Retrofitting Cities to Meet the Carbon Challenge*. Available from http://www.arup.com/_assets/_download/04EDBCEA-19BB-316E-40AE7DE4DE51E964.pdf, accessed 28 October 2008. See also http://www.arup.com/communications/feature.cfm?pageid=10239 for related text.

Head, P. (2008b) Entering the ecological age: the engineer's role. The Seventh Brunel International Lecture 2008, Desert Research Institute (4 October) Nevada, USA. Main web page http://www.arup.com/americas/event.cfm?pageid=11866 and lecture sides and notes from http://www.arup.com/_assets/_download/72B9BD7D-19BB-316E-40000ADE36037C13.pdf, accessed 28 October 2008.

Normile, D. (2008) China's living laboratory in urbanization. *Science* (319):5864 740–743. DOI: 10.1126/science.319.5864.740. http://www.sciencemag.org/cgi/content/full/sci;319/5864/740?maxtoshow=andHITS=10andhits=10andRESULTFORMAT=andfulltext=dongtanandsearchid=1andFIRSTINDEX=0andresourcetype=HWCIT.

Wood, R. (2007) *Planning Institute of Australia (PIA) 2007, Presentation on Dongtan.* http://www.arup.com/_assets/_download/8CFDEE1A-CC3E-EA1A-25FD80B2315B50FD.pdf, accessed 29 Aug 2008.

▨ FURTHER READING

Herbert Girardet (2006) *Creating Sustainable Cities* (Schumacher Briefing Series 2, Totnes: Green Books Ltd), gives an excellent and short overview of why we need sustainable cities, using London as an example for some of the relevant points about consumption, waste, and energy.

W. McDonough and M. Braungart (2002) *Cradle to Cradle (Remaking the Way we Make Things),* New York: North Point Books, summarizes the authors' vision of guiding principles in redesigning cities, furniture, shampoo, transport, and many other everyday items. They use their backgrounds in architecture, urban planning, and chemistry to consider new design frameworks which we can all use in order to become more sustainable.

▨ USEFUL WEBSITES

http://www.masdar.ae/home/index.aspx is the home page for the Masdar City initiative taking place near Abu Dhabi (United Arab Emirates), which will be very ambitious in terms of being zero-energy and zero-waste with an estimated development budget of US$ 22 billion.

http://www.arup.com/arup/feature.cfm?pageid=10322 is one of Arup's podcasts as part of their 'Drivers for Change' (foresight research programme) and includes educational talks on various topics such as demographics, energy (http://www.arup.com/_assets/_download/download630.pdf), and other interesting topics. Here Rachel Birch (http://www.arup.com/arup/newsitem.cfm?pageid=10317) talks about waste and the consumer impact on the throwaway society.

Finally, this site (Arup, again) gives Peter Head's vision on 'Entering the Ecological Age' http://www.arup.com/americas/event.cfm?pageid=11866; the framework used here is that there will be three ages: the agricultural age, the industrial age (which we are currently claimed to be in), and the forthcoming, hopeful ecological age.

22

Innovative partnerships for sustainable development in Harare, Zimbabwe

Charlene Hewat and Barbara Banda

Introduction

Sustainable development was initially defined by the World Commission on Environment and Development in 1987 as development that 'meets the needs of the present without compromising the ability of future generations to meet their own needs' (Brundtland 1987). This definition is now widely adopted worldwide.

Following the World Commission, the first world summit of national leaders on sustainable development (known as **UNCED**) was convened in Rio de Janeiro in 1992. It was followed 10 years later by the second world summit (**WSSD**) in Johannesburg, South Africa, which recognized the necessity of 'changing unsustainable patterns of consumption and production'. The 'plan of action', the main document to emerge from the second summit, called for 'fundamental changes in the way societies produce and consume' (Haas *et al.* 2002; see also Chapter 5 this volume where this is discussed in the East European context). It resolved to 'encourage and promote' programmes to accelerate the shift towards sustainable consumption and production in order to enhance social and economic development within the carrying capacity of **ecosystems**. This was to be achieved by meeting challenges of, and where appropriate de-linking, economic growth and environmental degradation through improving efficiency and sustainability in resource use and production processes, and reducing resource degradation, pollution and wastes (Haas *et al.* 2005).

> Carrying capacity is the supportable population of an ecosystem, given the food, habitat, water, and other necessities available within an environment.

The 2002 WSSD also advocated a major role for partnerships between **civil society**, government and business to work towards sustainable development. Such partnerships, involving clusters of firms in Harare, the capital city of Zimbabwe with a population of approximately 2.8 million in 2006, provide the context for this chapter which describes environmental management partnerships. The clusters are an initiative that provides a platform for a coordinated and holistic approach to sustainable environmental management by industry, local government, and civil society in the city. More importantly, the chapter discusses how the activities of these industrial clusters 'fit' within the broader goals of the sustainable development agenda. It examines how firms working together have responded to the environmental, developmental, and sustainability challenges facing their production lines and the services provided by the City of Harare. Industrial cluster areas are thus identified in this chapter as 'eco-friendly economic zones' and models of best practice reflecting a win–win scenario for development and environmental sustainability.

Most of all, however, the chapter illustrates what can still be achieved in a country which, throughout the early years of the twenty-first century, has been in a continuous deep economic, social, and political crisis.

Industrial clusters from an international perspective

Internationally, industrial clusters are mainly conceived as localized production systems, characterized by a number of small- and medium-sized firms which are involved at various phases of the production of a homogeneous product family, such as shoe manufacturing. These firms often specialize in similar aspects of the production process, for example leather technology, and integrate with each other through a complex network of relationships (Becattini quoted in Landstrom 2005; Porter 1998).

Industrial clusters have been classically studied in northern Italy, where their formation was driven by an interaction of economic, social, and cultural factors (Boari 2001). They are claimed to have had a significant positive impact on the development of the region, being conceptualized as self-regulating networks of capitalism, with shared values, knowledge, and strategies mainly focusing on productivity and efficiency. At the time of writing, there are around 200 industrial clusters in Italy, employing 2.2 million people and responsible for one-third of Italy's exports. In another example, the southern Indian state of Kerala, industrial clusters include research and development activities for continuous upgrading of technology (Kerala Bureau of Industrial Promotion, Cluster Tidings—Kerala, Newsletter May 2004).

The formation of industrial clusters in Harare

The industrial clusters within Harare comprise firms that are located in an easily defined physical area. Unlike the Italian clusters described above, however, these firms do not form a homogeneous product family—in fact they are quite distinct from one another. Critically, and also unlike industrial clusters elsewhere, they were formed initially because of one of the greatest environmental health challenges facing the City of Harare and its citizens: provision of sufficient, safe, clean water.

In 2005, five industrial clusters had been formed in Harare, four along the Mukuvisi River System and one in the outskirts of the city (Figure 22.1). One such is the Ardbennie industrial cluster, formed in August 2003. All companies within the cluster's defined geographic location, irrespective of size, process, or type of industry, are free to participate. There are about 20 companies located within the defined area and to date ten of these participate in the Ardbennie cluster activities (see Box 22.1).

Industrial activities and environment pressures go hand-in-hand as industry contributes to a major part of water, land, and air pollution within Harare, as is the case in many other cities globally. When Harare was first built by British colonialists in 1890, firms were initially situated along the main water system, the Mukuvisi River (Figure 22.1), which they used as the outlet for their entire waste disposal. The pollution levels along this river system have risen considerably over the years, to a point where legislative pressures and moral arguments are on most company board agendas.

The cluster concept in Harare grew specifically, however, out of a broader crisis with respect to water across the city. By the late 1990s the local authority's capacity to provide adequate water had become greatly reduced due to a combination of factors—poor governance, population growth, inadequate and old infrastructure, and rural to urban migration. All this was happening at a time when the country was, and continues to, experience its worst ever economic crisis.

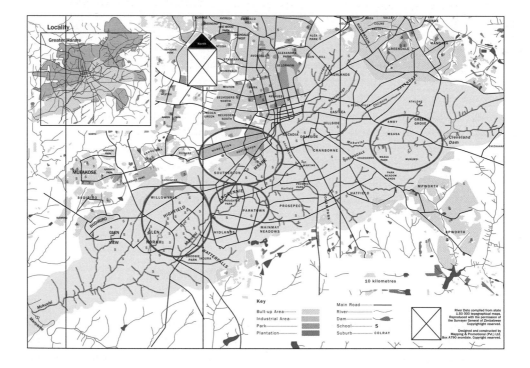

Figure 22.1 A map of Harare showing the four industrial clusters along the Mukuvisi River. These are circled, with the Ardbennie cluster in the smallest oval.

BOX 22.1 COMPANY PROFILES IN THE ARDBENNIE CLUSTER.

Company	Nature of business	Size of business
Nu Naks	Manufacture of snacks	Medium
Dunlop Zimbabwe	Warehousing and distribution of tyre products	Large
Pro Plastics	Manufacture of PVC piping and plumbing materials	Medium
Steel Line	Metal fabrication	Medium
Samburn Pressings	Tool and die making, metal pressings	Medium
Hamish Cameroon	Manufacture of stock feeds and oils	Medium
Waste Away	Collection and disposal of waste	Medium
Carrara Marble	Cutting, polishing, and installation of marble, granite and terrazzo	Medium
Astol	Haulage business	Medium
Cairns Foods	Manufacturing food and beverage products	Large

Alongside the other Harare industrial clusters, the Ardbennie Cluster participates in national and international events such as Wetland Day, World Water Day, World Biodiversity day and national such as Tree Planting Day, and occasional clean-ups of their area (see Figure below). Cluster members share best practices and improved technologies through exchange visits, reports, and regular meetings.

The Ardbennie Cluster organize a clean up within their area. This clean up involved The Deputy Mayor of Harare, the Permanent Secretary from the Ministry of Environment, and chief executive officers from the companies.

In response to the situation, the regional non-governmental organization, Environment Africa, organized a stakeholder-wide Water Crisis Conference in March 2003. The conference recommended the formation of a Water Task Force to spearhead the implementation of its resolution to find a lasting solution to the perennial water problems in the city and its dormitory towns. This Task Force, comprising representatives from

Environment Africa (civil society), The Business Council for Sustainable Development Zimbabwe (BCSDZ) (private sector), Harare City Council (local government), Zimbabwe National Water Authority (government parastatal), the Scientific and Industrial Research Development Council (SIRDC) (research institution), and the Property Owners Association, has worked on a number of initiatives with the broad objective of reducing water pollution and ultimately improving the quality and supply of water in the city. It was from this that the Harare Industrial Cluster concept was born where it was felt that industry, the major water user, could play a key role in addressing the crisis.

> Environment Africa is a southern African-based NGO that works with all sectors of society to protect and manage natural resources and to promote sustainable development, looking for African solutions to African challenges.

The industrial clusters in Harare present an opportunity, therefore, for companies within the same location to combine resources and work towards the improvement of their local environment and ultimately to contribute towards broader sustainable development. Although they arose directly out of the 2003 Water Crisis Conference, they could not have developed and been sustained without a number of background factors, or drivers, combining to create a force for change. These drivers comprised:

A paradigm shift in the City Council

For a long time, the City of Harare's approach to environmental issues followed a conventional top-down approach. It failed, however, to address the problems due to lack of funding and inability to monitor the situation. It was at this time that the industrial cluster approach was launched and the City Council was encouraged to revise its approach.

There had to be a shift in the City Council's conception of industry from the traditional view of regarding firms as the client paying for a service, to one of the private sector being part of the solution that contributes towards meeting environment and sustainability challenges. As a perceived neutral stakeholder, Environment Africa was able to bring together industry and the City Council to discuss and work on collective approaches that could be implemented at both individual firm and at the cluster level.

Legislative drivers

Zimbabwe is not only signatory to international environmental obligations such as the Kyoto protocol (on climate change) and Agenda 21 (on sustainable development) (Brook *et al.* 2002) (see Box 22.2) but has formed an Environment Management Agency (EMA) to address issues of environmental degradation. The international legislation to reduce carbon dioxide emissions, ban ozone-depleting substances, and protect biodiversity is being adopted by larger industries together with the adoption of voluntary international standards (such as ISO9001 and 14001) as well as the ISO26000 guide on (**corporate**) **social responsibility**. Clusters have enabled larger organizations to share and assist implementation of these international laws and voluntary standards with smaller members who otherwise would not know about them. The cluster approach has also given an opportunity for companies to participate in local policy

---

BOX 22.2 INTERNATIONAL ENVIRONMENTAL TREATIES, SCHEMES AND STRANDARDS

The Kyoto protocol of 1997 followed on from the UN Framework Convention on Climate Change which was a product of UNCED in 1992. It sets legally binding limits on signatory countries for emissions of greenhouse gases.

Agenda 21 was another output from UNCED. It introduced a comprehensive plan of action for the twenty-first century to further sustainable development at global, national, and local levels.

The International Organization for Standardization (ISO) establishes worldwide standards for many aspects of industry and commerce. The ISO 9000 series covers quality management systems and the ISO 14000 series sets environmental management standards. Companies can claim certified status if they meet the set standards. ISO 26000, due to be released in 2010, provides guidelines for social responsibility but, unlike the other two, is not a certification scheme.

---

decisions and some clusters have contributed to a new waste management policy. These initiatives thus contribute towards UN **Millennium Development Goals**, which seek to eradicate extreme poverty, ensure provision of health care and education, and ensure environmental sustainability.

Business drivers

Businesses exist to make profits where an important driver is the need to reduce operational costs as well as ensuring the availability of inputs, and this can be achieved through clustering. Water, being one of the main reasons for the establishment of the clusters, has been a major focus. Clusters have established and shared company water targets and measures to reduce water consumption. In most cases water usage has been reduced (see Figure 22.2) which in turn has reduced costs to the company. In one case a larger company within a cluster assisted a smaller company on the establishment of a water audit where the smaller company found burst pipes and has since reduced water costs significantly. As well as water, the clusters have also focused on waste management and developed exchange programmes where one company's waste becomes another company's input. Thus, they have developed joint collection and in some cases looked at opportunities to link their waste to small enterprises which recycle it for making alternative products. For example, wire offcuts from a company were provided to a microproducer who turned the wire into wire toys. The company has since turned its attention to conducting energy audits.

Social and environment drivers

Increasingly across the world, companies need to concern themselves with their images as socially responsible players, where a good image is also good for business. The Harare industrial clusters are able to demonstrate their commitment to public health for their employees and local residential communities by collectively cleaning up their act.

Broadly speaking, these cluster activities fall under the heading, **corporate social responsibility** (CSR). A specific example concerns Chloride Zimbabwe, a battery-manufacturing

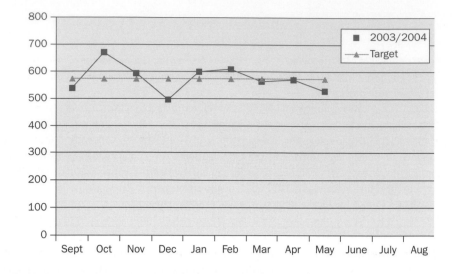

Figure 22.2 Hamish Cameron, a small company in the Ardbennie cluster: monthly reports on water audits and evidence of water saving for the year 2003–04.

company, which partnered Environment Africa to buy old vehicle batteries from local people for reuse and recycling. This has reduced the toxic waste in the environment, reused valuable raw materials which otherwise would be imported, provided livelihood opportunities for local people, and contributed towards the implementation of the Zimbabwe Government's Environment Management Act. It is an example of how the four pillars of sustainability—policy, social, economic, and biophysical (see Box 22.3) are working in tandem.

The drivers described above were all probably necessary for the formation of the Harare industrial clusters, but in themselves they would not have been sufficient. Of paramount importance is buy-in and support of the concept by the top management, ideally at chief executive or managing director levels. There is also need for a champion/facilitator—a company that will lead the process within each industrial area with the commitment of all the other companies involved.

Potential cluster impacts and challenges

Although impacts at the company level are easy to measure, sustainable development is a long-term goal and less amenable to measurement. The following medium- and long-term impacts are anticipated:

• Improved access to a reliable supply of water for local people.

• Improved energy efficiency within companies achieved by carrying out energy audits. With a national energy shortage, less energy used by industry allows more energy to be supplied to households.

BOX 22.3 THE FOUR PILLARS OF SUSTAINABILTY

This model considers sustainability as comprising four interactive and mutually reinforcing 'pillars' or dimensions:

1. Policy, which includes power and decision making;
2. Economic, which concerns jobs and money;
3. Social, which concerns people living together;
4. Biophysical, which concerns life, land, air and water.

For example, when a company decides to reduce and eventually stop polluting the river systems, some of the issues it will need to consider are:

- Policy issues: who is the legal authority, what is the legislation etc.?
- Economic issues: for example, how much is it going to cost to install a water treatment plant to clean waste water before disposal into the river system?
- Social issues: are employees or downstream communities being affected by polluted water from the firm?
- Biophysical issues: what life is being affected (e.g. the fish)? Is the land adjacent to the river being polluted?

Note that the model adds an extra 'pillar'—that of policy—to the three-pillar model often used. This is referred to in Chapter 1 and elsewhere in this book.

Adapted by Environment Africa from O'Donoghue (1995)

- Reduced water pollution through improved processes within industries. This will in turn reduce the chemicals being used by the City of Harare to treat the water. (Currently nine different chemicals, mostly imported, are in use). It will also help rejuvenate river systems, especially in terms of fish which is also a food source for city dwellers.

- Opportunity for poor, local communities through links with industry. For example, waste materials which would normally be thrown away are given to community groups to utilize for secondary products.

- Mainstreaming of environmental issues in companies.

It would, of course, be naïve to assume that industrial clusters around environmental issues represent a panacea for sustainable development. There are several challenges, some of them deep-seated:

- Some executive directors of firms perceive environmental and social issues as an add-on to their activities and don't provide the necessary leadership.

- CSR is understood in many cases as philanthropy (e.g. donating to charities). CSR needs to be understood as investment with a pay-off for business (see above).

- Cluster meetings take up valuable time, and hence are not prioritized. There is a need for mainstreaming of activities.

- The current political and economic instability within the country has impacted negatively on industrial development and although much has been achieved with the clusters to date, the national crisis has been a real constraint to their growth and maintenance.

There are, nevertheless, many possibilities for the future which build on the anticipated impacts while meeting those challenges over which there is some measure of control. These future possibilities can be described as follows.

Expansion of cluster activities

The majority of the cluster activities concentrate on environmental management with a focus on pollution abatement and, efficient water and energy use. While laudable, the next move is to expand the focus towards achieving broader sustainable development. Companies can be encouraged to adopt a life-cycle management approach towards more sustainable consumption and production patterns.

Broadening the participation base

There is also a need for cluster activities to build synergies with other sustainable development management programmes in Zimbabwe. This helps stimulate sharing of best practices and allows cluster activities to draw from the work being done on the broader environment, development, and sustainability agenda.

Extension to other sectors

The incorporation of specialized capacities, for example from universities, technology licensing organizations, and financing institutions into cluster activities would help offer new opportunities.

Replication

The industrial cluster initiative is an example of an innovative approach to sustainable development and partnership that can be replicated not only within Harare but in other parts of Zimbabwe and the region. A similar model has been established in Victoria Falls with assistance from Environment Africa. Here, the tourism industry has organized itself to improve its overall management and minimize its impacts on this sensitive, biologically diverse area, which is a World Heritage Site.

■ SUMMARY

- The 2002 World Summit on Sustainable Development (WSSD) identified government–private sector–civil society partnerships as an important mechanism for delivering its agenda. The Harare industrial clusters provide an example of such partnerships in action.

- Industrial clusters have historically been viewed in terms of promoting economic development. The Harare clusters are different in that they are specifically a response to a pressing social and environmental health issue—the need for clean drinking water in the city. They are also a response to the need for reliable water supplies for industrial processes.

- Firms in the clusters try to achieve win–win situations, meaning that activities that are good for the wider community (such as improved quality of, and access to, water) are also good for business.

- Some cluster firms have broadened their activities to include environmentally friendly waste management practices, again on a win–win basis where possible.

- A major challenge for sustaining the Harare clusters is support from top management in firms.

- Opportunities to scale-up the cluster concept in a number of ways are emerging.

■ REFERENCES

Boari, C. (2001) *Industrial Clusters, Focal Firms, and Economic Dynamism: A Perspective from Italy*. Available from http://sitesresources.worldbank.org/WBI/Resources/wbi37186.pdf. Accessed 10 March 2009.

Brook, T., Mittermeir, R. A., Mittermeier, C. G., da Fonseca, G., Rylands, A., Konstant, W., Flick, P., Pilgrim, J., Oldfield, S., Magin, F. and Hamilton-Taylor, C. (2002) Habitat loss and extinction in the hotspots of bio-diversity. *Conservation Biology* 16(4): 909–923. Quoted in Population Dynamics and MDG7 an essay offered as a discussion document for the Population-Environment Research Network (PERN) cyberseminar from 5–16 September 2005. Available from http://www.populationenvironmentresearch.org/seminars.jsp. Accessed 10 March 2009.

Haas, W. (2002) What do Social Systems Consume? A different view on Sustainable Consumption. In *Interim Report IR-02-073, Lifecycle Approaches to Sustainable Consumption workshop Proceedings*, 22 November, 2002. E. G. Hertwich (ed.). Austria: International Institute for Applied Systems Analysis, Laxenburg.

Landstrom, H. (2005) *Pioneers in Entrepreneurship and Small Business Research*. New York: Springer.

Porter, M. (1998) Clusters and the new economics of competition. *Harvard Business Review* November December: 77 90.

23

Community-led sustainable development: the champion approach

Charlene Hewat and Barbara Banda

Introduction

Until the late 1980s most development work could be described as 'top-down' or 'core-periphery'. This approach was based on the understanding that investments can be made in specific sectors of the economy or geographical areas and the benefits will spread and help other areas. The 'top' constituted government, which created development policies and projects without consultation with local people. Most failed in one way or another. The reasons for failure were many and often specific to a particular policy or project. A general reason that has endured, however, is that they did not involve the supposed beneficiaries in their design and implementation, and hence did not understand the local contexts, and did not enjoy local buy-in.

From the 1980s, participatory approaches to development policy and practice have made significant inroads, where an influential champion has been Robert Chambers (1997). Such approaches are now fairly mainstream, with many international agencies promoting them, including the World Bank in its book *From many lands* (Narayan and Petesch 2002).

This chapter is written by two members of Environment Africa, a regional NGO which is based in Zimbabwe (see also Chapter 22), and draws on our personal experience to explore a variant of participatory approaches. Called 'community-led development', the variant is about developing active and sustainable communities, which involve community leaders, the community itself, activists, non-governmental organizations (NGOs), government departments, and professionals working together to improve livelihoods.

Environment Africa has taken up community-led development with farmers in rural areas of the country, but with a particular angle that focuses on key players or 'champions'. It is based on the following maxim:

Go to the people, live with them, and learn from them, and love them. Start with what they know. Build with what they have and work with the champions. When the work is done, the task accomplished, the people will say, we have done this ourselves.

Lao Tsu (700 BC)

The champion approach in this case can be defined as an informal system of 'best practices extension' whereby progressive, committed, and passionate local farmers are identified by Environment Africa, with assistance from community and local government extension officers. These champion farmers are encouraged to share and pass on knowledge and inputs, such as technical know-how, seed and/or livestock, to fellow farmers. The champion approach is akin to the farmer-to-farmer extension approach rather than the traditional dominant top-down approach of extension officers who mostly promote rigid and prescriptive ideas (see Box 23.1).

Once the champion farmer has been identified they are trained and provided with basic inputs for selected interventions, such as banana plantations, small grains production, intercropping methods, livestock etc. The vision, however, is not just about improving immediate livelihoods, but of sustainable development with its focus on protecting the environment in order to meet the needs of future generations. Where poverty is prevalent, environmental protection is understandably often not a priority for communities and the role of Environment Africa is to teach the farmers its importance in their farming practices if they are to have sustainable livelihoods. This is not always an easy task and, although there is an appreciation of the importance of the environment, many times there is little sustainable use because the immediate aim of such communities is survival today, with little thought about the future.

In this chapter we tell the story of Environment Africa's engagement with one such champion, which demonstrates how the approach in a community-based project in the Zvimba area of Zimbabwe (see map, Figure 23.1) has expanded the capacities of the local community and improved their livelihoods economically and materially, environmentally, and socially. In keeping with the bottom-up approach, which targets individual champions as change agents for the whole community, it is based on a case study about a single family.

BOX 23.1 AGRICULTURAL EXTENSION

Farmer-to-farmer extension means a farmer passing on their knowledge and skills to fellow farmers, who in turn extend their knowledge and skills onwards to others. As local farmers themselves, they are able to achieve synergies between abstract ideas about best practice and their knowledge of the local context. Farmer-to-farmer extension contrasts with the traditional form of agricultural extension whereby agricultural professionals transmit specialist knowledge (sometimes emanating directly from agricultural research stations) to farmers, often without exploring innovations already existing in the community.

Figure 23.1 Map of Zimbabwe showing the location of Zvimba.

The Mubaiwa family: from subsistence to champion farmers

Mr Douglas Mubaiwa together with his wife Taneta and family come from a rural smallholder farming community in Zvimba District. Mr Mubaiwa is a subsistence farmer, relying on dry land cropping and hiring of cattle for ploughing. He is committed to hard work, growing citrus fruit and bananas, but in the past lacked technical knowledge and therefore remained static and at times failed to produce a meaningful harvest. The 8-hectare farm had its own challenges of erosion which resulted in deep gullies caused by poor farm practices and poor sandy soil conditions. Water was very scarce in the area. Due to poor capacity, related to lack of knowledge and material resources such as cattle to assist with the ploughing, only 2 hectares of the farm area were under cultivation. In 1992, Mr Mubaiwa received basic training in farming by the local government agricultural extension office which improved his skills in gully reclamation and soil conservation. Then, ten years later in 2002, he became aware of Environment Africa and was interested in what we had to offer.

Below we offer two perspectives on what happened next—one from Mr Mubaiwa himself, and one from Environment Africa (i.e. ourselves, the chapter authors)

In his own words: how Mr Mubaiwa became a champion farmer

We live in a rural area in two huts and don't even have a toilet. To plough our fields we had no cattle to help us pull the draught power so we hired these from our neighbours. Our fields were infertile and [erosion] gullies had developed everywhere. Around our homestead we had five banana plants and a few mango and guava trees. Water was a big problem and to water the plants and trees we dug a trench to try and lead run-off water towards them. Involving our family members, we decided to carry out an audit of our homestead and fields so as to help us plan. This was a difficult exercise and raised many conflicts within the family especially around the roles and responsibilities of family members. We worked with each others' strengths and tried to find ways to harmonize our differences and came up with a simple plan of action.

When Environment Africa came along they acknowledged the commitment and hard work that had already been carried out on our 8 hectare farm, especially the way we tried to improve our soils. To do this we used the leaf mould from the nearby forest and dug this in around our plants. The plan we developed was improved with the help of Environment Africa and together we came up with a Family Action Plan (FAP) which involved all members of the family. The FAP is now our guide on how to move forward in a holistic way and improve our farming methods, increase our yields, and better the way we live.

Environment Africa invited us to a number of training workshops on soil conservation and rain water harvesting at a micro-level and how to use an 'A' frame to trap rain water. They also taught us about conservation farming methods and soil improvement techniques. We diversified our crops and started intercropping our bananas with cassava, rapoko, cow peas, and other crops appropriate to dry weather. Our neighbours began to get jealous and would let their livestock in to destroy our crops at times. This made us think about how we could involve our neighbours in our plan and we requested Environment Africa to extend training to them. Our advisors from Environment Africa were thrilled with this idea and more village meetings, field days, and commemorations were organized for the surrounding farmers. I was also given a bicycle by Environment Africa so that I could move around and share my knowledge and experiences to other distant farmers. Our homestead also became a classroom where other farmers would come to learn. Today we are a very proud family with a herd of 14 cattle, chickens, turkeys, and an improved homestead with a toilet.

The others think that makes us proud and we are a model in our community and beyond. We also have various families in the neighbourhood that have joined our Environment Action Group (EAG), and have followed in our footsteps and are successful. Today we have over 120 members in the EAG and are grateful to Environment Africa for their training and small inputs, such as fencing and seed. The other thing that makes us proud is that we don't require any food aid. We are also receiving international visitors from Europe and locally we are referred to as the environmental champion.

The Environment Africa perspective on the champion approach with the Mubaiwa family

The entry point into communities to establish the champion approach is very important as each has a different history, socio-economic profile, culture, and traditions. We were welcomed by the Mubaiwas and together carried out a scoping and mapping exercise using standard participatory techniques so as to find out what was working, the challenges, and conflicts.

We started off with the provision of technical assistance in, for example, sustainable water harvesting techniques so as to improve the water supply and soils. For this, we taught them how to use the basic 'A' frame tool. An 'A' frame is made up of three pieces of wood or bamboo joined together to form the frame, with a string hanging from the apex with a stone attached to the end. The stone swings and centres itself to give a line reading on the ground of where contours can be placed. The A-frame can also be used to make level terraces to collect and hold run-off water as well as achieve proper drainage of a field. We also supplied the family with fencing materials and basic start-up inputs such as seed.

Water-conservation farming as described above was just one of the overall techniques in which we trained the family and it was interesting to learn how they integrated their indigenous knowledge into the programmes. For example, once we visited the smallholding and the family showed us how they kept their bananas watered throughout the dry season by means of the water melons that they grew in-between the banana plants. Family members punctured a hole in the top of each water melon. They then dug a hole next to the banana plants, turned the watermelon upside down and covered it with dry grass and soil. This, they said, was natural drip irrigation. This was new and amazing to us, and is something that we now include in Environment Africa training.

It didn't take long before the Mubaiwa family began to realize visible benefits such as achieving food security and income from their produce and they have built up their own herd of 14 cattle from the two they bought with the initial income they earned. They have also improved their homestead and they now take care of their grandchildren who are orphaned. It was only when the Mubaiwa family started earning income from this project that they started sending their grandchildren to school.

Intercropping between the bananas with beans and groundnuts has helped to revive soil fertility through nitrogen fixation. (Beans, groundnuts, and other plants of the legume family use nitrogen from the air and, by the action of bacteria in their roots, convert it or 'fix' it into a soluble form in the soil that is then available as a plant nutrient.) Other crops such as cassava, maize, rapoko, and grain amaranth are also included in a crop rotation and, as part of their crop diversification efforts, the Mubaiwas introduced herbs and useful trees such as the Moringa tree, which is edible and highly nutritious. This tree, commonly known as the horseradish tree, is native to India. It grows quickly from seed and the community can eat the leaves as relish, and the green pods. The seeds can also be crushed into oil and used for cooking. Although none of the herbs or trees grown is a cure for the devastating HIV/AIDS pandemic that is prevalent throughout Zimbabwe, they can improve the nutritional diet and thus improve the immune system.

Members of the community were attracted to the family's achievements and began requesting help for their own areas. It was at this stage that Mr Mubaiwa championed the formation of an Environment Action Group (EAG) in Zvimba, with interested members from the community and technical assistance from Environment Africa.

The Zvimba EAG now has its own constitution and leadership structure, which is chaired by Mr Mubaiwa himself. The EAG was launched with an initial membership of seven in 2003, which had grown to 139 (70 males and 69 females) in 2007. The interest shown by women is largely due to the realization that issues addressed by the EAG are of immediate benefit to them as they are the main providers of food, water, and energy for their households. The awareness provided by Environment Africa on sustainable use of these resources is key. For example, simple things like washing in the rivers contribute towards water pollution and Environment Africa has shown them how to minimize this by washing on the side of, rather than in, the rivers.

The EAG promotes farmer-to-farmer sharing and learning visits as well as produce fairs. The fairs enable farmers to come together, to exchange seed, skills, and knowledge as well as give each other encouragement. In his testimony Mr Mubaiwa says, 'My life has been transformed, my family can sit down and share ideas around natural agricultural and other everyday social issues.'

We also remember one of our Environment Africa visits where Mr and Mrs Mubaiwa were carrying out a farmer-to-farmer training and he pulled us aside and said, 'Look over there in the distance, that is my forest and I am going to protect it and use it for beekeeping.'

It was then that we realized that our own emphasis on sustainable development had been seriously understood, appreciated, and implemented in what we call 'the African way'. This is a way that is inclusive and is embedded in cultural norms and behaviours. An example is the use of sacred places, where plants and animals are controlled by the traditional norms and behaviours of a community. These traditional norms and behaviours are respected by the community more than the imposed rules.

This is a story of a true champion whose legacy will be used by future generations. This is a sustainable land use model which is being replicated in other parts of Africa. We are now using the champion approach throughout Zimbabwe, and have recently introduced it into neighbouring countries such as Malawi and Zambia.

The Mubaiwa story is one of many success stories that we have to share and learn from. The champion approach identifies and trains an individual who has passion, is quick to adapt and able to share their experience and best practices with the rest of the community. The main philosophy about the champion approach is that of a Chinese proverb, 'give a man a fish and you feed him for a day. Teach a man to fish and you feed him for a lifetime.'

Lessons learned from the Mubaiwa family champions

The above case study illustrates the economic, social and environmental dimensions of sustainability and their interrelationships when related to people's livelihoods. Aided by Environment Africa capacity building and investment in basic infrastructure such as fencing, the Mubaiwa family and other families in Zvimba have become economically self-sufficient and well able to meet their needs. Environmentally, these families have adopted conservation farming methods such as water harvesting, intercropping, and composting which has led to improved soils and therefore improved productivity, and water supply. The champions took up the initiative to remove alien, invasive tree species such as *Lantana camara* which inhibits the growth of other plants, poisons cattle, and destroys pasture lands and indigenous tree species. They have also adopted reforestation programmes, which have led to increased water supply, and in turn improved the yields as well as allowing vegetable gardens all year round.

Sustainable economic and environmental benefits could not have come about, however, without a deepening of social ties within and between families, facilitated by Environment Africa's participatory methodologies. For example, the Mubaiwa family has been brought closer together, by creating processes, such as the Family Action Plan, which help develop a collective vision and resolve conflicts. These social dimensions can also be seen between families in the community, where social networks have been strengthened through agreeing joint responsibilities and supporting each other via the Environment Action Group (EAG). Environment Africa's role has been to act as a broker for building relationships and to pace the processes and hence allow time to build trust between members through communication and reflection.

More broadly, relationships have been developed between communities, the local Agriculture Extension Office, the local municipality, and the Forestry Commission. Through listening to the community some problems beyond Environment Africa's terms of reference in the area emerged and Environment Africa has linked the community to other organizations with the capacity to deal with issues such as HIV/AIDS. In the absence of alternative institutional mechanisms, the EAGs have become an important resource base and a social security system for the community.

The results have been encouraging. For example:

- Improved access to water, reducing the burden on women and ensuring that they have time for other duties.

- Improved life skills capacities and knowledge base.

- Diversification of crops has provided the community with food security and nutritional security through improved meal frequency and balanced diets.

- Dependency on outsiders, for example aid-providing NGOs, has been reduced.

- Because there are visible results, new members are attracted to the EAG who then replicate the practices. In other words, the results are scaleable.

There are of course challenges to be overcome. A significant time investment is required, especially at the start. The selection of champions can be seen as privileging a few individuals which can give rise to community jealousies. The capacity building/training budgets for Environment Africa are high, and understanding from donors who fund Environment Africa is required. It is important too that the selected champions understand that this approach is about capacity building, not only of them as individuals, but of the whole community, and it therefore carries responsibilities. Also, it certainly is not about providing them with finance. In this last respect, a related undermining issue concerns the continued, but ultimately unsustainable, presence of other NGOs in the area who provide communities with huge inputs upfront before understanding whether or not they have the capacities to use them.

Despite these challenges, the overall impact has been positive. The champions have become true change agents throughout the community and the EAG, and a focus for discussion, sharing, and learning. In conclusion, we can see an empowered community in the making, exhibiting a collective energy for change.

▓ SUMMARY

- Community-led development involves participatory methods which involve the farmers themselves in agricultural development and associated environmental conservation.

- Participatory methods draw on farmers' own knowledge of local contexts. Because of this it is argued that they produce better results than top-down agricultural extension work, which makes use primarily of professional knowledge with limited sense of application to local conditions.

- The chapter reports on community-led development in a rural area of Zimbabwe, where the regional NGO Environment Africa has applied what it terms the 'champion approach'. The approach involves identifying local farmer champions with whom the NGO works to build up first the farmer's own capacities. This provides an exemplary model which the farmer then shares with other members of the local community.

▓ ACKNOWLEDGEMENTS

The authors would like to thank all the staff at Environment Africa who helped us put this chapter together, and the community of Zvimba. A very special thank you to the Mubaiwa family, you are our inspiration.

▓ REFERENCES

Chambers, R. (1997) *Whose Reality Counts? Putting the First Last.* London; Intermediate Technology Publications.

Lao Tsu (700 BC) www.anecdote.com.au/archives/2008/02/lao_tsu_on_comm.html, accessed February 2009.

Narayan D. and Petesch P. (2002) *From Many Lands. World Bank 'Voices of the Poor' Series.* Washington, DC: World Bank.

▓ USEFUL WEBSITES

http://www.eldis.org/go/about-eldis. ELDIS is a knowledge service provided by the Institute of Development Studies, University of Sussex, UK, which aims to share the best in development policy, practice, and research.

24

Why involve the public?

Case studies of public involvement in environmental initiatives in South Africa

Kevin Winter

Introduction

Since the early 1980s community-driven environmental action has gained increasing recognition (Warburton 1998). During this time researchers attempted to understand why people get involved in environmental actions, how public interest in such actions is retained, and how public involvement can be made more meaningful (Stukas and Dunlap 2002).

But why is it a good idea to involve the public in the first place in environmental actions? A common answer assumes that the public are in the best position to understand local conditions and therefore are likely to provide solutions appropriate to local circumstances (Carr and Halvorsen 2001). Some researchers even claim that public participation is a pre-requisite for sustainable development, peace, and social justice (Brown and Quiblier 1994). However, while there is growing evidence of public participation in decision making and in public actions affecting the local environment (Warburton 1998; Stukas and Dunlap 2002), not all researchers are as enthusiastic about the outcome. Questions are being asked about the risk of involving the public in actions that might have long-term consequences, for example, in situations where the public takes responsibility for managing scarce resources (Hauck and Sowman 2003).

This chapter explores public involvement in three environmental initiatives at Betty's Bay, a small holiday town situated on the south western Cape, South Africa. 'Involvement' here refers to individuals or groups of individuals who engage voluntarily in some form of social activity or action to achieve certain goals (Stukas and Dunlap 2002). Typically this involvement is initiated and occurs in a social organization, such as a committee or an association of people, and is characterized by planned, pro-social behaviour over a reasonably long period of time (Dobson 2003).

The three environmental initiatives are explored as separate case studies, but taken together they focus on the potential of voluntary interest-based groups to contribute to environmental actions. The earliest began in 1962 and involved the clearance of alien vegetation. The second initiative started in 1972 as the Wild Flower Society, and later the Botanical Society. In 2000, a third initiative was established that focused on the problem of water pollution in one of the coastal lakes at Betty's Bay.

The geographic context of Betty's Bay

Betty's Bay is a small coastal township that stretches 11 kilometres along the southern Cape coast of South Africa (Figure 24.1). The town is approximately 100 kilometres from Cape Town city centre.

The township of Betty's Bay is one of four urban settlements situated within the United Nations Educational, Scientific and Cultural Organization (UNESCO) Kogelberg Biosphere Reserve (KBR). The reserve has some of the most significant remnants of the Cape Floral Kingdom, one of the six floral kingdoms of the world. Indigenous plants cover an area of almost 90 000 km² in this region (Cowling and Richardson 1995).

The reserve is bounded to the west and south by a rugged coastline of sandstone rock, interspersed occasionally by sandy bays and beaches. Only a narrow coastal plain separates this coastline from the foothills of the mountain peaks beyond. Streams flowing from the mountain slopes are impounded by coastal sand dunes and are responsible for the formation of three black-water acid lakes (Boucher 1978), including Bass Lake (Figure 24.2)

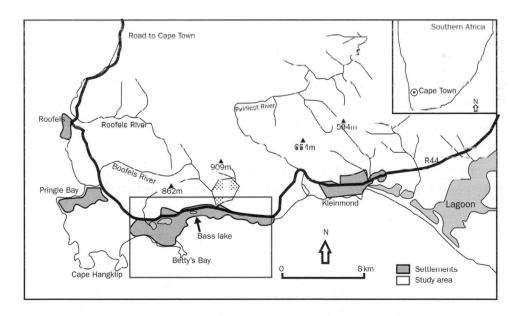

Figure 24.1 Kogelberg Biosphere Reserve with the study area shown by the rectangle.

Figure 24.2 Steep mountain slopes of the Kogelberg in the background, and a narrow coastal plain with Bass Lake in the foreground.

which is one of the case studies below. These lakes represent some of the best examples of rare, black-water acidic habitats in South Africa (Johns and Johns 2001).

The acidic state of the lake water is caused by the surrounding sandy soils and the release of tannins from the roots of natural vegetation. The dark colour of the water is also the result of these tannins flowing into the lake.

Betty's Bay has expanded significantly over the past 20 years. In 1987 the township was connected to the national electricity grid system, thereby offering an attractive service to existing residents and potential buyers of land and property. Later in 1996 the coastal road was upgraded so as to reduce significantly the travel time from Cape Town to Betty's Bay and make it more accessible to weekend holidaymakers than ever before.

South Africa's colonial and apartheid history is characterized by social conflict and injustice. Legislation during the apartheid era forced the separation of racial groupings and this had a profound effect on the social milieu of Betty's Bay. White South Africans benefitted by acquiring land and property along the southern Cape coastline in a racially segregated area. In general, those that did purchase property at Betty's Bay were relatively affluent, well-educated, often professional people, with sufficient leisure time and capital to enjoy weekends away from the city. The social mix remains largely unchanged despite a change in government. In 2001 over 90 per cent of the permanent residents of Betty's Bay were white South Africans (Statistics South Africa 2003).

Perspectives on collective action

Collective actions involve a number of people carrying out the same or similar actions at the same time in which there is some form of mutual dependency among the participants (Coleman 1994), but is collective action more effective in achieving sustainable development than individual action? History suggests that public action is more likely to lead to the transformation of social conventions, norms, and institutions. Challenging the status quo is more likely to succeed by agents acting collectively rather than single agents acting alone (Sztompka 1991). Seippel (2002) suggests that the only remedies and hope for addressing seriously degradation of the natural environment lies in voluntary social action and cooperative public involvement. There is plenty of evidence over the past two decades, at least in the United Kingdom and United States, to indicate a raised level of public participation in decision making processes and public actions affecting the local environment (Warburton 1998; Stukas and Dunlap 2002).

The following explains four overlapping assumptions which are used to explore the potential of individuals to act as a collective:

1. Individuals do not necessarily act exclusively on the basis of self-interest.

2. Individuals can be expected to cooperate and act collegially.

3. Collective actions are informed by local knowledge and understanding of local conditions.

4. Individuals often act collectively by accepting norms and values of a group of people.

Individuals do act beyond self-interest

Individual actions do not necessarily occur as a result of an individual acting at will, but rather that the behaviour of others can and does influence individuals to act along a particular course of action (Gilbert 1989). Action can therefore be thought of as being 'social' in that it takes the behaviour of others into account.

There is nothing new in this idea that individuals are capable of behaviour that extends beyond their individual self-interests. In 1937, Talcott Parsons, a social theorist, explained how individual actions interconnected with other individual actions resulting in a complex web of social interaction called an 'action system'. He argued that individuals could be expected to establish shared goals through this system.

The ideas of Parsons may not appear to be immediately obvious in a discussion on sustainable development. However, inherent in the concept is the idea of going beyond self-interest through acting in ways which do not jeopardize the needs of future generations. Through interaction and mutual agreement, therefore, it can be expected that sustainable development will become increasingly acceptable as an idea that can be put into practice.

Cooperation in collective action

While shared goals can be developed through social interaction, they in turn lead to cooperative actions that are likely to have mutually beneficial outcomes (Miller 2001). This implies that cooperation is based on an incentive in which individuals expect to benefit in some way. Parsons (1937) claimed that cooperation is underpinned by an acceptance and adherence to norms and values that are accepted and adhered to in a stable, orderly society (Seippel 2002). He also stated that a failure to adhere to norms and values would lead to disruption, anarchy, chaos, and aimlessness (Parsons 1937). A state of agreement and cooperation is initiated and maintained through a process of socialization in which individuals and groups accept the prevailing or perceived norms. Without socialization, the potential to achieve shared ends is limited because conflict prevents agreement from being reached. Parsons is criticized for taking this position because he seems to deny the possibility that society can strengthen its resolve and seek direction through conflict (Mullard and Spicker 1998). South Africa is a good case in point in that it shifted from a position of conflict resulting from racial intolerance to a state of democracy.

Actions that reflect local knowledge and understanding

Collective actions in local environmental issues usually assume that the public is in the best position to understand local conditions and is therefore more likely to provide appropriate solutions (Carr and Halvorsen 2001). O'Riordan and Stoll-Kleemann (2002) claim that a transition to sustainability will be achieved partly through a participatory democracy in which global issues are understood at a local scale, and where local actions are linked to the global. This interplay between these two scales reflects a growing realization that globalization is a process that impacts on places, resources, and livelihoods worldwide. A recurring message of Agenda 21, the key document that emerged from the United Nations Conference on Environment and Development (**UNCED**) in 1992, is that sustainable development initiatives are likely to be most effective at a local scale involving ordinary people.

Actions based on collective norms and values

The contemporary understanding of social capital—the ability of people to associate and work together for common purposes (Coleman 1994)—provides new insight into the acceptance of norms and values by collectives. Social capital is influenced by social interaction and communication, networks, relations of trust, knowledge, norms, and values (Coleman 1994; Etzioni 1996). Well-developed social capital is linked to a strong internal morality in which individuals balance their individual rights with collective responsibility. Collective responsibility appears to be closely bound to an acceptance of norms and values. Moral order then rests on core values that are largely shared by society and are embedded in its social structures (Etzioni 1996). Shared values are considered to be those values to which citizens are committed to upholding. Human behaviour is partially explained by understanding the nature of the norms that are being followed. For example, a prescriptive norm that obliges an individual to engage in taking collective responsibility might encourage that person to forgo their self-interests and to act in the interests of the collective (Coleman 1994).

Societies often have strong moral voices that help to maintain social order in which values are accepted voluntarily rather than being forced to do so (Etzioni 1996). A 'moral voice' should therefore encourage individuals and groups to reflect on their shared values and to avoid behaviour that contributes to unsustainable development.

The three environmental initiatives of Betty's Bay

The assumptions discussed above can now be used to inform the three selected case studies.

Control of alien vegetation

In the early 1950s Hangklip Beach Estates Company exercised its right to develop the township of Betty's Bay. In order to stabilize the movement of the coastal dune system, the company introduced several alien plants and trees. Soon thereafter exotic and alien species, such as the Australian eucalyptus and black wattle trees, were introduced by private landowners to create hedgerows, windbreaks, and shaded areas (Macdonald *et al*. 1985).

Alien vegetation spread rapidly beyond its intended domains and soon invaded the indigenous vegetation. Concerns about this encroachment were first raised at an Annual General Meeting of Betty's Bay and District Ratepayers Association. Mr Heesom, a Cape Town journalist and property owner at Betty's Bay, proposed that volunteers should take responsibility for systematically removing these invasive alien plants. He acknowledged that the task would require a concerted effort and commitment. He therefore invited volunteers to meet on the first Sunday of each month at a predetermined site to commence the battle. He championed this initiative for over sixteen years until his death in 1979. Subsequently the leadership of the Hack, as it became known, was passed onto members of the Betty's Bay Wild Flower Society, which later became known as the Kogelberg Botanical Society (see below).

The Betty's Bay Hack group is a good example of a long-term commitment and resolve to address the invasion of alien vegetation and to reduce its impact on local floral diversity and beauty. The Hack group has always maintained that growth of alien vegetation on public land will never be completely prevented because seeds are easily carried by wind, insects, and birds from private land. In other words, the removal of alien vegetation would require ongoing involvement of volunteers. While local by-laws were passed to prevent private land owners from planting alien vegetation, this proved difficult to enforce. Nevertheless significant gains were achieved. Today nearly all public land is free of alien vegetation. Attwell (1994: 10) adds that the 'dedication and efficiency of Hack convenors together with the loyalty and hard work of the "troops" established a camaraderie unequalled in any other sphere of Betty's Bay life'.

Wild Flower and Botanical Society

After a devastating fire in February 1970, wild flowers sprung up all over Betty's Bay during the spring of 1972. These circumstances were largely responsible for the formation of the Wild Flower Society (Attwell 1994), established later that year by a small group of permanent and weekend residents. The Wild Flower Society became known as the Betty's Bay Botanical Society in 1986, and later still, the Kogelberg Botanical Society in 1992.

Flower shows were arranged by the Society with the explicit intention of emphasizing the conservation of indigenous vegetation and providing an education for property owners. The flower shows also drew attention to the floral diversity at Betty's Bay, and raised awareness of biodiversity conservation (Attwell 1994).

The wide interests and initiatives of the Society also provided an opportunity for members to became actively involved in a variety of other environmental initiatives including efforts to protect African penguins that arrived unexpectedly on the mainland at Betty's Bay, the continued coordination of the Betty's Bay Hack Group (see above), and support to the development of the Kogelberg Biosphere Reserve.

Sewage and water initiative

In 1984 the Municipality admitted that poor management of sewage at Betty's Bay had the potential to pollute the freshwater lakes. By January 1998 water pollution became increasingly evident. Local residents and visitors who had used one of the lakes for swimming and boating complained that the water smelt of 'rotting vegetables'. The lake was overloaded with nutrients. Water samples showed that *Escherichia coli* (*E.coli*) bacteria exceeded the South African Water Quality Guidelines acceptable for human contact (DWAF 1996).

> *E.coli* is a bacterium commonly found in the lower intestine of warm-blooded animals. Most *E.coli* strains are harmless but some strains can cause diarrhoea and even serious food poisoning in humans.

In 1999 a committee was formed by local ratepayers. Then in 2000 an algal bloom in one of the lakes was found to contain toxic blue–green algae. Algal blooms should not exist in these lakes because of their acidity (Gardiner 1988). It was assumed that the pollution originated from failed septic tank systems (small household domestic sanitation systems) on properties surrounding the lake.

The local authority was slow to react but eventually agreed a request from the committee to fund researchers from the University of Cape Town so they could conduct a low-cost monitoring programme. Among other things, the study provided a motivation for the local authority to draft a policy to prevent home owners from installing septic tank systems in areas with a high water table. The study also identified sites where individual tanks were failing. In these cases the systems were replaced at the local authority's expense.

The sewage and water initiative resulted in the general improvement in water quality at Bass Lake. The committee continues to research alternative sewerage technologies with the intention of finding a sustainable solution.

Learning from the case studies: collective action in practice

The three initiatives described highlight common features. The dominant aim of each initiative focuses on the conservation, protection, and improvement of the biophysical environment. Secondary aims are found in efforts to inform and educate the public and ratepayers (landowners).

In particular, they each illustrate the four assumptions about collective behaviour (Table 24.1). In all three initiatives there is evidence to show that volunteers worked

Table 24.1 Characteristics of collective behaviour associated with each initiative.

Criteria	Alien vegetation campaign	Wild Flower and Botanical Society	Sewage and water initiative
Does involvement contribute toward the common good?	Inform, educate, and communicate with ratepayers Conserve heritage of public land	Educate public and ratepayers Conserve plant biodiversity on public land Efforts to establish a nature reserve	Inform and educate ratepayers Search for alternative sewage disposal Improve the management of the public lake
Is there evidence of cooperation?	Camaraderie Volunteer support for collective action (the Hack)	Collective, voluntary effort to prepare flower shows	Collaborative effort of committee in discussion, decisions, and actions
Is local knowledge being developed?	Acquired knowledge of alien plants and their management in the local environment	Acquired specialist knowledge of indigenous (and alien) vegetation	Local knowledge used to identify sites for monitoring Acquired specialist knowledge from researchers and applied to local environment
Is there evidence of accepting collective norms and values? (evidenced by taking joint responsibility).	Took responsibility for: • The Hack as a regular exercise • Educating property owners • Planting indigenous vegetation	Took responsibility for: • Preparation of flower shows • Lobbying local authority to conserve African penguin colony • Supporting the Botanical gardens • Educating public and property owners	Took responsibility for: • Establishing and ensuring ongoing monitoring programme • Seeking funding • Educating and informing residents • Searching for alternative sewerage systems (ongoing)

cooperatively together over a reasonable period of time. Without such cooperation, it is unlikely that these interest groups could have achieved any measure of success.

All three interest groups demonstrate goodwill and a commitment to the goals of the respective initiatives. In general these acts of goodwill have contributed to the common good by conserving public open spaces. These claims are substantiated by evidence that shows:

1. The conservation of floral biodiversity through the systematic removal of alien vegetation over a period of 40 years;

2. An awareness among ratepayers of the need to preserve and protect indigenous flora at Betty's Bay;

3. The establishment of a penguin reserve that later resulted in the development of a public amenity;

4. Joint cooperation between ratepayers, local authorities and researchers to improve the water quality of a lake so as to ensure its continued use as a public recreational amenity.

Conclusions

All three initiatives were started because individuals identified an environmental problem or issue at Betty's Bay. In each case, public interest in the local environment led to the identification of issues worthy of the attention and involvement of volunteers. The sustained effort of interest groups over a reasonable period of time is characterized by cooperation, camaraderie, and a sense of purpose.

The case study identifies at least five reasons why the involvement of collectives might contribute to sustainability:

- Interest groups can be expected to establish goals for the common good of the public and conservation of the environment.

- Interest groups collectively build local knowledge, skills, and expertise that may otherwise be unavailable to individuals or local institutions had they acted alone.

- Membership of an organization offers greater opportunity for a group to access a range of opportunities, including 'outside' assistance and expertise, and financial support.

- Interest groups can be expected to take collective responsibility to address public interests and aspirations.

- Interest-based groups are capable of achieving modest and incrementally valuable results over time as demonstrated by the outcomes of these initiatives.

The initiatives at Betty's Bay shows how a group of people, who enjoyed a measure of socio-economic and political advantage as a result of the apartheid political system, have managed to work together over a reasonable period of time. Now that the country has shifted radically towards a democratic political ideology, the challenge is to extend the same social assumptions and thinking discussed in this chapter to other areas in the country particularly where a large proportion of the population are poor and marginalized. In these areas, where service delivery is weak and natural resources are degraded, environmentally-related self-help typically concerns public health, such as initiatives to implement low-cost water and sanitation technologies (Carden *et al.* 2007).

▓ SUMMARY

- Involving the public can contribute towards sustainable development. Three case studies from the Cape area of South Africa illustrate that, through collective public action, individuals are likely to:

  (a) Act beyond self-interest, establish common goals based on shared norms and values, and take collective responsibility for environmental concerns.

  (b) Contribute their local knowledge and apply specialized knowledge to the local consequences of global environmental concerns.

▓ REFERENCES

Attwell, J. (1994) *A History of the Betty's Bay Wild Flower Society.* Kleinmond: Ekkoprint.

Boucher, C. (1978) Cape Hangklip area II: the vegetation. *Journal of Bothalia* 12(3): 455–497.

Brown, N. and Quiblier, P. (1994) *Ethics and Agenda 21: Moral Implications of a Global Consensus.* New York: United Nations.

Carden, K., Armitage, N., Sichone, S. and Winter, K. (2007) The use and disposal of greywater in the non-sewered areas of South Africa: Paper 2—greywater management options. *Water SA* 33(4): 433–432.

Carr, D. and Halvorsen, K. (2001) An evaluation of three democratic, community-based approaches to citizen participation: surveys, conversations with community groups, and community dinners. *Society and Natural Resources* 14: 107–126.

Coleman, J. (1994) *Foundations of Social Theory*. Cambridge, MA: Harvard University Press.

Cowling, R. and Richardson, D. (1995) *Fynbos: South Africa's Unique Floral Kingdom*. Cape Town: Vlaeberg.

DWAF (Department of Water Affairs and Forestry) (1996) *South African Water Quality Guidelines*. Pretoria: DWAF.

Dobson, A. (2003) *Citizenship and the Environment*. Oxford: Oxford University Press.

Etzioni, A. (1996) *The New Golden Rule: Community and Morality in a Democratic Society*. New York: Basic Books.

Gardiner, A. (1988) A Study on the Water chemistry and plankton in blackwater lakelets of the south-western Cape. Unpublished PhD thesis, University of Cape Town.

Gilbert, M. (1989) *On Social Facts*. Princeton, NJ: University Press.

Hauck, M. and Sowman, M. (2003) *Waves of Change*. Cape Town: University of Cape Town Press.

Johns, A. and Johns, M. (2001) *Kogelberg Biosphere Reserve: Heart of the Cape Flora*. Cape Town: Struik.

Macdonald, I., Jarman, M. and Beeston, P. (1985) *Management of Invasive Alien Plants in the Fynbos Biome, South African National Scientific Programmes Report No. 11*. Pretoria: CSIR.

Miller, S. (2001) *Social Action: a Teleological Account*. Cambridge: Cambridge University Press.

Mullard, M. and Spicker, P. (1998) *Social Policy in a Changing Society*. London: Routledge.

O'Riordan, T. and Stoll-Kleemann, S. (2002) *Biodiversity, Sustainability and Human Communities: Protecting Beyond the Protected*. Cambridge: Cambridge University Press.

Parsons, T. (1937) *The Structure of Social Action*. New York: McGraw Hill.

Seippel, O. (2002) Modernity, politics, and the environment: a theoretical perspective. In R. Dunlap *et al.* (eds), *Sociological Theory and the Environment*, pp. 197–229. Lanham, MD: Rowman and Littlefield Publishers.

Statistics South Africa (2003) *Census 2001*. Pretoria: Department of Statistics.

Stukas, A. and Dunlap, M. (2002) Community involvement: theoretical approaches and educational initiatives. *Journal of Social Issues* 58(3): 411–427.

Sztompka, P. (1991) *Society in Action: the Theory of Social Becoming*. Chicago, IL: University of Chicago Press.

Warburton, D. (1998) A passionate dialogue: community and sustainable development. In D. Warburton (ed.), *Community and Sustainable Development: Participation in the Future*, pp. 1–39. London: Earthscan.

■ **USEFUL WEBSITES**

http://www.saiea.com/calabash: the Calabash project seeks to support democratic reform through participatory decision making by providing appropriate tools, knowledge, and networks to regulators, civil society, practitioners, and industry of the Southern African Development Region (SADC) region.

http://www.iclei.org: Local Governments for Sustainability (ICLEI) is an international association of local governments as well as national and regional local government organizations that have made a commitment to sustainable development. This site features many case studies on community participation in environmental initiatives and is regularly updated with contributions from all over the world.

25

Tools and techniques for environmental decision making

Pamela Furniss and Chris Blackmore

Introduction

Making decisions is something that we all do every day, individually and in groups, at work, in our communities, and at home. But what makes a decision, an environmental decision? And why do we need tools to make them?

Environmental decisions can be defined as any decisions that have an effect on our environment. They range from national level decisions such as a government deciding to expand air travel, to decisions by a company to invest in renewable energy technology through to an individual's decision to reduce and recycle household waste. They include all decisions that have an environmental dimension and, in the context of this book, acknowledge that such decisions have developmental and sustainability dimensions as well.

Decision making lies at the core of many of the case studies described in this book. For example in Chapter 12, The National Park Authority in Scotland, with members representing a number of different interests, has to reach decisions that balance conservation, recreation, and local livelihood goals. In Chapter 21, the designers of Dongtan have to integrate environmental, social, economic, and cultural factors into their planning decisions. In Chapter 22, the stakeholders in the industrial clusters of Harare reach a collective agreement on (among other things) the clean-up of the local river.

In all these cases, and for all but the simplest decisions, there are many different factors to be taken into account. Environmental decision making (EDM) means including environmental considerations alongside many others—economic, social, political, and technological.

The very nature of EDM processes means that they can often be characterized by needs to:

- collect and analyse large amounts of data in many different forms,
- identify and involve key stakeholders in decision making,
- take multiple perspectives on decision situations into account,
- identify the purpose of the decision process, and
- monitor and evaluate the outcomes and consequences.

These characteristics provide the rationale for using tools and techniques. These diverse tasks are frequently highly complex and challenging, making the use of tools and techniques essential to help understand and compare alternative ways forward. Many different factors influence decision making and there are many different approaches to EDM, involving different degrees of rationality. Hence a wide range of tools and techniques suitable for different aspects of EDM is required. Some of them are discussed in the next section.

Tools and techniques

To many of us a tool is something like a hammer or a tin opener. In the context of EDM, tools and techniques can be anything from physical tools (like pen and paper, tape measure, pH meter), to computer software, to large-scale methods and processes (English 1999). Different tools can be used for different purposes within an EDM context. For example, some tools would be used for data gathering (both quantitative and qualitative), some for data analysis (both conceptual and mathematical), some for thinking, for action, for communication, for prediction, for involving people in the decision-making process, and many more.

Within the scope of this chapter, there isn't sufficient space to include a comprehensive description of all the many possible tools and techniques that could be used for EDM, but Table 25.1 provides a summary of a selected few. (For further details of these and several other tools and techniques, see *Techniques for Environmental Decision Making* (T863) and other sources in Further reading.)

This selected list comprises tools and techniques that are used at different stages and relate to different levels of decision making. We refer to several of them below when we demonstrate the framework for decision making. Some, such as data gathering and brainstorming, are used in the early stages and can be applicable to many situations. Diagramming and modelling include a wide range of techniques, which can have value for understanding and communicating at all stages of a decision process. Others, such as decision trees and multi-criteria analysis, have more specific application and formalized techniques such as EIA are applicable to specific developments. Participation potentially underpins all of them if decision making is conceptualized as an inclusive process.

Table 25.1 Selected tools and techniques for environmental decision making.

Technique	What it is	Useful for
Backcasting	Thinking and planning process, for use by individuals or groups, in which you identify your desired future situation and then plan the decisions and action stages required to get there.	Working out how best to achieve an imagined future situation; makes you identify goals and purposes.
Brainstorming	Free thinking and ideas generating activity, best done in a group, in which people offer up their thoughts on an issue.	Exploring a situation by getting issues out in the open, from all participants. Needs a second stage of grouping and rationalizing the ideas or you can end up with a long, unmanageable list.

continued

Table 25.1 (Continued)

Technique	What it is	Useful for
Cost–benefit analysis (CBA)	An economic model that uses monetary valuations of the costs and benefits associated with a particular decision or series of decisions and, by comparing costs against benefits, reduces the data to a single figure for each possible decision.	Presents a numerical valuation that enables simple comparison between different courses of action. Criticized because it requires assigning monetary values to intangibles such as environmental quality, loss of amenity etc. or may ignore the intangibles altogether.
Data gathering	Includes a wide range of techniques from simple physical measurement (e.g. land area), through questionnaires and surveys to maps and geographical information systems (GIS) software.	Provides essential information to enable understanding of the situation. Depends on having the appropriate resources of time, money, equipment, skills etc.
Decision trees	Technique for analysing a decision by breaking it down into sub-decisions, estimating probability and ascribing values to different outcomes in order to choose between them.	Clarifying possible options for decision making by giving them a numerical value. Subjective process relying on the judgement of the person(s) doing it.
Diagramming	Representation of a situation in a diagram that may include a combination of components, arrows, shapes of various types etc. There are many types of diagram each with particular conventions and guidelines for how to draw them e.g. systems maps, rich pictures, multiple cause diagrams, influence diagrams, sign graphs etc.	Exploring and analysing a situation to reveal and communicate different understandings and relationships. Different types of diagram will portray different aspects of a situation.
Impact assessment	Formal procedures that are legal and/or regulatory requirements in several countries. Includes EIA (environmental impact assessment), SEA (strategic environmental assessment), sustainability appraisal, risk assessment, SIA (social impact assessment), and others.	Assessing and predicting effects of major projects. Value can be questioned if not all perspectives are included or if objectivity of assessors is in any doubt.
Environmental management systems (EMS)	Practical framework for managing environmental issues and implementation of environmental policy within a company or organization. ISO 14001, the international specification for an EMS, is widely used.	Structured approach should ensure consideration and minimization of all potential impacts. Formal certification can bring commercial credibility on environmental matters.
Modelling	Many different types of modelling are applicable to EDM. For example, mathematical computer-based models can be used to predict possible outcomes based on input data. 3D models can be used to represent landscapes or development projects.	Useful for dealing with large amounts of quantitative data. Dependent on the modeller(s)' understanding of the situation, its components and the relationships between them. Providing a simple, visual demonstration of possible future situations.
Multi-criteria analysis	Ranking technique for identifying a single, most-preferred option from several complex choices.	Can be useful in a situation where there are several variables to consider. Very dependent on the judgement of the people doing the analysis.
Participation	Group of techniques that are relevant to all types of decision making. They include workshops, focus groups, citizens' juries, and others.	Involving stakeholders in the decision making process. Can be time-consuming. Needs to be planned and enacted with care to be effective. Not to be confused with consultation, which is used to gain stakeholder views on possible actions, but which excludes them from the decision making process.
Stakeholder analysis	At its simplest, a list of stakeholders which can be grouped if interests are shared or overlap (stakeholder mapping).	Understanding the scope of the situation and appreciating the range of interests that may need to be considered. An important element of many other techniques.

Evolution of EDM tools and techniques

When did people start using EDM tools and techniques and why? How have they changed over time? Are they likely to change very much in future? The history of these tools and techniques can be tracked back a long way. Modelling, in terms of making representations of situations to explore understanding and communicate with others, has been used for centuries. The earliest environmental legislation and ways of implementing it were probably developed in Roman times and concerned reasonable use of water. Cost–benefit analysis, in terms of weighing up the positive and negative elements of a project, originated in the late nineteenth century. During the 1930s, this developed into a formal requirement to compare costs and benefits in the context of water-related investments in the western USA (Pearce 2006). By the 1950s cost–benefit analysis had become an established economic process with consistent methods. It was frequently adopted for large-scale projects, for example, in the UK it was the standard method of assessment for transport systems such as the development of new roads.

However, cost–benefit analysis and similar techniques did not, at that time, usually take environmental issues into account. The development of tools and techniques specifically for environmental decision making has been much more recent. During the 1960s and 1970s the need to incorporate environmental assessment and management was increasingly realized and reflected in environmental legislation. Environmental impact assessment (EIA) was first introduced in the USA as a result of the National Environmental Policy Act of 1969 and today, much of the world has at least partial regulations for EIA. EIA provides a formal process for assessing the environmental impacts of a project. Not all EIA processes are the same but typically they consist of a number of stages including data collection, prediction and evaluation of environmental impacts, consultation, preparation of a written environmental impact statement and monitoring and evaluation of the effects of the project. EIA is now standard practice in many parts of the world but its value can be questioned if not all perspectives are included. For example, EIAs for major development projects such as large dams have sometimes been criticized for inadequate public consultation, especially of the people directly affected or displaced by the project. There may also be different interpretations of what is meant by the 'environment' and therefore what is included within the scope of the EIA. Displacement of affected people may be seen by some as a social rather than environmental impact and therefore omitted from the process.

Since the mid-1970s, several other environmental and sustainability assessment techniques have been developed as the importance of including social, as well as environmental, dimensions became increasingly apparent. Social impact assessment (SIA) and risk analysis were incorporated into environmental assessment during the late 1970s (Rose et al. 1995). There were also changes of level as well as scope. Although originally applied to site-based projects, EIA evolved to apply to policies, plans, and programmes at strategic decision making level. Strategic environmental assessment (SEA) was the subject of a European Union (EU) Directive in 2001 and incorporated environmental and social issues into development planning and decision making. In the UK, SEA is closely allied with Sustainability Appraisal, an assessment of the economic, environmental, and social effects of a development plan to ensure it is considered to be sustainable development.

Less formal techniques have also evolved over time, sometimes being adapted from one context to another. For instance, some participatory techniques have a long history in international development contexts and have spread to more explicitly environmental contexts as the close relationship between environment and development has become more apparent. In recent times in Europe use of participatory techniques for facilitating multi-stakeholder decision making processes in environmental contexts has become more formalized since the introduction of the Aarhus Convention (adopted in 1998 and enforced since 2001). This is the United Nations Economic Commission for Europe's Convention of Access to Information, Public Participation in Decision Making and Access to Justice on Environmental Matters. The Aarhus Convention requires that public participation is an integral part of decision making processes by EU states.

Another factor that has affected the evolution of tools and techniques has been advances in technology. For example, advances in computing capacity have enabled large quantities of data to be handled more easily for modelling of complex environmental processes and scenarios. Examples can be found from contexts of climate change, e.g. the modelling of impacts, adaptability, and vulnerability that have been publicized through the Intergovernmental Panel on Climate Change process. Capacities and abilities to map, measure, and communicate have also increased remarkably with advances in technology, e.g. increased use of the internet for gaining access to information, use of email and mobile phones in some kinds of data gathering, and use of digital instruments for monitoring and measuring water and air quality.

Development of tools and techniques has also come about because of the way they have been used. Tools such as geographical information systems (GIS) can be used quite mechanistically with an emphasis on collecting data. However, GIS can also be, and has been, used much more creatively, for instance in enabling stakeholders to find out about an area together, influencing their thinking about, say, managing their water resources in the process of using the tool (SLIM 2004a).

Looking to the future evolution of tools and techniques for EDM, it seems probable that technology will continue to develop. It's possible this could increasingly mask some of the assumptions of modellers in mapping and measuring because fewer background 'workings' need be shown with more 'user-friendly' machines. To counter balance this trend, innovation in the use of tools and techniques (in the sense of the GIS example above) also looks set to continue. The huge amount of environmental legislation may well continue to grow but there is also increasing recognition of the need for other ways of encouraging change, such as using incentives and social learning as well as the enforcement and guidance of legislation. Incentives can be financial, cultural, and ecological e.g. evident as grants or cost-cutting, approval of others and enhanced environmental quality. Social learning involves interactions among stakeholders in situations at different levels of governance to bring about a range of improvements. According to the SLIM water management project, processes of social learning that can be observed include:

> the convergence of goals, criteria and knowledge, leading to more accurate mutual expectations, and the building of relations of trust and respect . . . the emergence of agreements on concerted action; the process of co-creating the knowledge needed to understand issues and practices; a change in behaviours, norms and procedures arising from development of mutual understanding of the issues, as a result of shared actions such as physical experiments, joint fact-finding and participatory interpretation.

> (SLIM 2004b: 2)

SLIM (2004b) also highlights that such interactions among interdependent stakeholders require proper facilitation, institutional support, and a conducive policy environment. Tools and techniques have an important role in all three of these areas, among them various 'frameworks for decision making' that focus on how the many different processes discussed above can work together. One such framework is discussed in the next section.

A framework for decision making

Table 25.1 highlights some specific tools that each have their own particular purpose. How could these tools fit into an environmental decision making situation? Figure 25.1 shows a possible framework, or suggested process, for making environmental (or other) decisions.

To 'read' this diagram, start at bottom-left entering the process by exploring the situation. Appropriate tools to use here might be stakeholder analysis, brainstorming, or some sort of diagramming. Then move on to formulate problems and opportunities—this might involve modelling or some stakeholder participation methods. From there, you can identify feasible and desirable changes—on a large scale this might require EIA or CBA or at a more local level, backcasting or multi-criteria analysis. Having decided what needs to be done, you can it put into practice by taking action. Actions will have consequences so the process doesn't finish there, you continue to monitor and evaluate, or re-explore, the situation. You will notice in the centre of Figure 25.1 the rectangle represents the

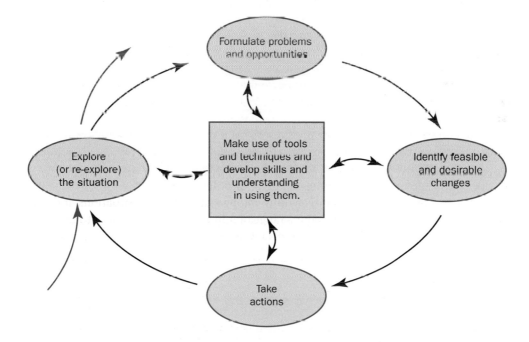

Figure 25.1 A framework for environmental decision making. Adapted from Open University (2006).

tools and techniques that can be used throughout the process. The double-headed arrows to and from the central rectangle indicate that the framework is not intended to be used rigidly as a step-by-step process, but that you use the tools as needed at any of the four stages and can skip stages or go back and forth between stages. The framework is not a prescriptive direction of how to make decisions, rather it is itself a tool to aid decision making and should be used flexibly and creatively.

Conclusion

This chapter has given an overview of the range of tools and techniques that can be used to aid environmental, developmental, and sustainable decision making. We have only been able to outline a few possible approaches—you will need to look elsewhere to learn more about the details of these techniques. It also needs to be said that tools alone are of little value unless you have skills in using them. If I was presented with a plumber's tool-box, that wouldn't make me a plumber—I wouldn't have the necessary skills in how to use the tools. As Figure 25.1 points out, you need to use the tools by putting them into practice and thereby develop skills and understanding before they can really aid the process.

Some of the techniques outlined here are fairly straightforward and can be adopted in many situations. They don't all require much in the way of resources other than the people to participate and the time and space to gather together and follow the process. Others are more demanding in terms of expertise, money, and support services and so could not be easily undertaken by a lay person.

Choosing the right tool therefore not only requires awareness of what you're trying to achieve but also of the resources available to you. When selecting a tool, you need to consider the pros and cons in terms of time required, human and financial resources, expertise, complexity, and most importantly whether it is the right tool for the job. It's also worth remembering that tools can be used in different ways—even if you have the right tool for the job, it can be used poorly or perhaps, in the context in which it is used, it might not be the appropriate tool.

In any context the desired outcome is a successful decision. But who judges success? This is likely to differ for different stakeholders. Politicians, developers, community groups, affected individuals are all likely to have different views and it may not always be possible to please all stakeholders all of the time. However, ultimately if their views are considered and integrated effectively into the process then the outcome has a much better chance of success.

■ **SUMMARY**

- Environmental decision making involves diverse tasks that are frequently complex and challenging, making tools and techniques essential for understanding and comparing possible ways forward.

- There are many different tools and techniques that can be used in environmental decision making, each with a particular purpose and value.

- These tools and techniques have developed over time in response to the increasing importance of environmental and sustainability issues.

- A framework for environmental decision making, applicable to many different situations, is presented.

■ REFERENCES

English, M. R. (1999) Environmental decision making by organizations: choosing the right tools. In K. Sexton, A. A. Marcus, K. W. Easter and T. D. Burkhardt (eds), *Better Environmental Decisions*, pp. 57–75. Washington, DC and Covelo, CA: Island Press.

Open University (2006) *T863 Environmental decision making: a systems approach*. Milton Keynes: The Open University.

Pearce, D. W., Atkinson, G. and Mourato, S. (2006) *Cost–benefit Analysis and the Environment: Recent Developments*. Paris: OECD Publishing. Executive summary available from http://www.oecd.org/dataoecd/37/53/36190261.pdf.

Rose D., Dalal-Clayton, B. and Hughes, R. (1995) *A Directory of Impact Assessment Guidelines*. London: IIED.

SLIM (2004a) *Facilitation in Policy Processes: Developing New Professional Skills SLIM Policy Briefing No 4*. Available from http://slim.open.ac.uk/objects/public/slim-pb4-final-screen.pdf, accessed Dec 2008.

SLIM (2004b) *The Role of Learning Processes in Integrated Catchment Management and the Sustainable Use of Water. SLIM Policy Briefing, No 6*. Available from http://slim.open.ac.uk/objects/public/slim-pb6-final-screen.pdf, accessed Dec 2008.

■ FURTHER READING

Open University (2006) *Environmental decision making: a systems approach (T863)*. Milton Keynes: Open University.

T863 course texts:

   Techniques for environmental decision making

   Book 1: Introducing environmental decision making

   Book 2: Starting off systemically in environmental decision making

   Book 3: Making environmental decisions and learning from them

   Book 4: Critical appraisal in environmental decision making

All are available from http://www.ouw.co.uk/bin/ouwsdll.dll?COURSET863_Environment. See also http://www3.open.ac.uk/courses/bin/p12.dll?C02T863

Annie Donnelly, D. Barry Dalal-Clayton, Ross Hughes, International Institute for Environment and Development, World Resources Institute, International Union for Conservation of Nature and Natural Resources (1998) *A Directory of Impact Assessment Guidelines*. London: IIED.

Blackmore, C., Ison, R.L. and Jiggins, J. (eds) (2007) Social learning: an alternative policy instrument for managing in the context of Europe's water. Special issue of *Environmental Science and Policy* 10(6).

Virginia H. Dale and Mary R English (1999) *Tools to Aid Environmental Decision Making*. New York: Springer-Verlag.

■ USEFUL WEBSITES

For details of the Aarhus Convention: http://ec.europa.eu/environment/aarhus/

International Institute for Environment and Development (IIED): http://www.iied.org/

Institute of Environmental Management and Assessment (IEMA): http://www.iema.net/

Intergovernmental Panel on Climate Change (IPCC): http://www.ipcc.ch/

Section D review

Section D has amply illustrated the third premise of this book—that there are many views on sustainability in relation to environment and development and, especially for this section, many views on how to achieve it. Thus, in this section we see themes concerning:

- policy and how policy is arrived at
- effective technologies and innovations (and their relationship to policy processes)
- desirable processes, and associated 'tools', for negotiating and enacting shared meanings of sustainability.

Any notion of action for environment, development, and sustainability will be a composite of desirability and feasibility. The latter represents what is or might be possible given our knowledge of the 'facts' about the world. The former concerns our values, or what we think should be the case.

Partly the chapters in Section D explore (or assume) what constitutes good action, and partly they examine the related question, 'Who is to act?' One answer to the latter might be government (or governments acting together at the international level), as surely government is the only body which can legitimately represent large groups of people, even if such representation appears inadequate. Alan Thomas (Chapter 17) examines the issues in terms of the government of a small country—Wales in the United Kingdom—and we soon note that smallness does not mean simple. Wales is one of the few nations with sustainable development enshrined in its constitution, but this has arisen at least in part through the pressures exerted by a number of non-governmental actors who continue to play a major role in influencing public policy. We cannot isolate government action, therefore, from other influences, a process which we call **governance**. A further issue concerns institutionalization and scaling up from individual projects. A sustainable development policy at the top of government does not necessarily mean it automatically becomes institutionalized in the everyday practices of departments and related agencies.

Godfrey Boyle (Chapter 18) also considers government actors but at the level of cities—in this case London. The city authority appears keen to be a major player among global cities in effecting major reductions in carbon emissions and the chapter highlights evolving policies that are supported by successive Mayors. In contrast to Thomas' emphasis on the multiple actors involved in policy and institutional change towards sustainable development, Boyle's contribution is to consider the potential of technology to provide solutions, and the politics involved when making technological choices.

Issues of policy process also feature in the chapter by Joanna Chataway, Peter Robbins, and James Smith (Chapter 19). The chapter illustrates how scientific, technological and economic ideas, and corresponding action, around 'tissue-culture' bananas in Kenya were institutionalized through various communications processes and recruiting a wide range of actors—including research institutes, international aid donors, and the small-scale farmers who would grow the bananas—to the cause.

Chataway, Robbins, and Smith discuss the issues in terms of a broad view of innovation as putting knowledge (from numerous sources) to productive use. More usually, however, innovation is viewed as new products and processes arising from capitalist markets (which are not well-formed in Kenya). Robin Roy (Chapter 20) also takes up the theme of innovation from the general, but different, angle of the processes and products that are necessary to achieve major reductions in energy and resource use, and in emissions and waste, up to 2050 and beyond. He argues that there is no shortage of individual 'green design' products but they are not sufficient, and it is necessary to design at the sustainable innovations level, by which he means 'developing whole systems that are environmentally, socially, and economically sustainable'.

When considering the actors who will 'do it', we should not neglect the private sector which has classically been the source of innovation in terms of shaping and responding to markets. In relation to environmental damage, the private sector has been seen as the bad guy because it is the prime agent for economic growth which, as we have seen in previous sections, is often in tension with environmental concerns. James Warren's description of the work of Arup, the design consultant engineers who hope to create an eco-city near Shanghai, China (Chapter 21), shows how private sector companies can be seen as part of the solution as they seek win–wins where corporate social responsibility is also good for business. The Chinese eco-city moreover can be seen as an example of sustainable innovation in the Roy sense of the previous chapter.

Chapter 22 by Charlene Hewat and Barbara Banda also sees the private sector as part of the solution to environmental problems in Harare, the capital city of Zimbabwe, where it acts in partnership with local communities, local government, and a regional environmental NGO. Chapter 23 by the same authors is again about Zimbabwe, where the emphasis is on diffusion of low-tech innovations for sustainable development in rural areas (although the term 'innovation' is not used in the chapter). Together, the chapters show hope and possibility despite the crisis in that country.

Chapters 22 and 23 also concern collective, participatory action at a local level, espousing participation as good action. The theme is taken up by Kevin Winter (Chapter 24) as he switches country to Zimbabwe's neighbour, South Africa. Taken together, these three chapters explore at a more local scale than is usual in the previous chapters the importance of collective and holistic actions which are based on a co-production of knowledge through participatory processes. They also illustrate ways in which the 'developed' north might learn good action from the 'underdeveloped' south.

Section C ended with a chapter whose frameworks provided thinking tools about how we might use ethics analysis to make sense of the interplay between facts and values in environmental decisions. In this section, tools to help arrive at appropriate trade-offs and decisions also form the basis of Chapter 21 on Arup, and are at least implicit in the three chapters involving Zimbabwe and South Africa. Meanwhile, **life cycle analysis** (or assessment) appears as a general tool in Chapters 20–22. This is an interesting point about all action—we develop tools to help us act together and think collectively about what to do. The final chapter in Section D (Chapter 23) by Pamela Furniss and Chris Blackmore takes up this general point about tools for environmental decision making. They outline some of the more important and widely used ones, and how, like physical tools, potential users need to know the best one(s) for a particular job and also how to use them well.

If we characterize Section D as being as being as much about co-production of knowledge for innovation and action as it is about the actions and innovations themselves, we return full circle to the basic themes of this book—difference and interdependency. As in the fourth premise from Chapter 1, we learn from each other and hence create new knowledge out of our differences. We are prepared to learn from each other because we are, in the final analysis, interdependent in everything we do. Although perhaps counter-intuitive, successful participation, partnerships, alliances—anything that implies acting together—are predicated on being able to learn from our differences even when that can be an uncomfortable experience, while recognizing a basic interdependence which necessitates working collectively. The difficulty remains, however, as signalled in the fourth premise and previous sections—difference is not necessarily a positive attribute as it is usually also based on inequality and social power differentials between individuals and groups. We turn to these issues in the concluding Section E.

SECTION E

SECTION E
Conclusion

26

Scientific, social science, and technological approaches to understanding environmental change

David Cooke (Technology), Michael Gillman (Science), David Humphreys (Social Science), and Gordon Wilson.

Introduction

Over the centuries human beings have evolved particular approaches, or specialisms, for analysing phenomena. Called 'disciplines', they 'discipline' people into thinking and analysing in particular ways according to different bodies of knowledge and intellectual traditions. Economics, political science, geography, and sociology in the social sciences; chemistry, physics, and biology in the natural sciences; systems, design, and engineering in technology, are all disciplines that are relevant to the study of environmental change. Each has added greatly to our knowledge. The problem, however, is that a real world phenomenon such as environmental change does not fit easily into any single discipline, hence an interdisciplinary perspective encompassing several disciplines is necessary.

All of the authors in this book have had some form of university education, including the practitioner authors. They are therefore likely to have been trained in one or more of the disciplines and they have, without necessarily explicitly stating it, brought disciplinary perspectives to bear in their analysis. For example, Chapter 2 on China uses an economics disciplinary perspective, while Chapter 11 on biodiversity and rainforest management is informed by the discipline of ecology and Chapter 20 on designing for sustainability by the discipline of design.

This book has introduced a range of disciplinary perspectives, in keeping with its emphasis of embracing difference. It's useful, therefore, to be able to identify from which discipline(s) writers are coming when thinking about their contributions, and also why their contributions are important. This chapter helps you to do this. With insufficient space to cover all disciplines, the chapter covers three broad categories of discipline—natural science, social science, and technology. Each has a separate author.

Why is understanding the science of environmental change essential?

Think about an old-fashioned clock. You can observe the second hand go round and may occasionally see the minute hand change. If you look again after a few minutes you will see that the minute hand has moved but it may take several more minutes before you see any difference in the hour hand. The hands of the clock are, in part, analogous to environmental change. Different changes operate at different rates and therefore require different observation periods to record change. An important distinction between the hands of a clock and environmental change is that the former proceeds at a steady pace whereas the latter may stop and start or fluctuate, often in an unpredictable manner. However, just as with a clock, environmental change, be it the numbers of fish in the sea, the levels of mercury in a river, or the amount of carbon dioxide in the atmosphere, can be monitored without reference to any underlying mechanisms.

In this analogy, the cogs and springs of the clock are equivalent to the processes driving environmental change. Why study these processes—the science of environmental change? One answer is obvious: without a knowledge of the underlying mechanism of environmental change humans can never understand how to influence that change. Thus, as a clock slows, removing the back might reveal a spring which had become unwound and needed winding, or an accumulation of dirt which was slowing the movements. Similarly, environmental change may have multiple underlying causes. Some of these may be more amenable to human control than others. A second related answer is that knowledge of the processes driving environmental change is important for predicting future change. However, past change is not necessarily an indicator of future change. If it were then all we need to do would be to record change as carefully as possible and then extrapolate to any future point using some appropriate mathematical or statistical model.

Often the science is less conspicuous and serves either as a basis for commentaries on socio-economic perspectives or as a background to technological innovation. An example of the first case is the 'ecologically sustainable farming skills' discussed in Chapter 16 on ethics. The phrase 'ecologically sustainable' may seem straightforward but, as Chapter 11 shows, it is underpinned by a large body of scientific endeavour. Biodiversity falls into the same category; an example is Chapter 19 which discusses the use of tissue culture. As the authors point out, the scientific method may seem relatively simple, but it is only so because much scientific effort has been put into perfecting it, for example, by understanding how cells interact.

A further subtlety in the science of environmental change, and science in general, is not only that there is a foundation of scientific results, but that they also employ scientific method. Such method includes objective observation and recording, replication of study, and experimental design. Experimental removal of deer and its effects on regeneration of plants is discussed in Chapter 12. In this case we do know about the design of the experiment, but even the simplest of this type of field experiment would allow a contrast and statistical analysis of the effects of low deer numbers compared with high (normal) deer numbers. (A field experiment is one that is undertaken outside under natural conditions rather than in a laboratory—field experiments do not have to occur in fields!) Management or legislation decisions based on such experiments will be more robust than decisions guided by intuition or observation of change over time, e.g. numbers of deer over time.

There are some cases of environmental change when field experiments are not possible. However, there are usually viable scientific alternatives. Whilst large-scale experiments of climate change are not possible, we have seen how scientific interpretation has yielded understanding of the mechanisms of climate change. Small-scale microcosm experiments are possible to simulate the effects on plants, e.g. looking for growth responses under different carbon dioxide concentrations. Indeed, we are currently undertaking an unplanned and uncontrolled experiment at the largest spatial scale possible through release of greenhouse gases into the atmosphere. This experiment, although grand in scale, is poor in its conception, with no control and replication and little opportunity for early cessation.

The role of the social sciences

As Michael Gillman has argued, science plays an important role in identifying the underlying causes of environmental problems. On some issues, such as marine pollution and climate change, the scientific debate is now settled, at least on the main causal relations. So why can't the politicians agree? Why have many of the responses to environmental problems been so inadequate to date? To answer these questions we need to look to the social sciences.

Social scientists seek to theorize on the reasons why social change occurs, and how it should be understood. They are essentially concerned with the social world: with collectivities of people and groups in society, the similarities and differences between them, the interdependencies that draw people together, and the various boundaries that come between them. For example, political theorists study the contestation between different actors over the making of decisions in government, society, and the international political system. Economists focus on both macro-economic factors, such as economic growth, employment and unemployment, and micro-economic factors, such as the decisions of individual consumers and firms. Geographers are concerned with the differences across time and space and the interconnectedness across scales. They point out that the causes of any space's characteristics do not lie exclusively within that space; so action in one space may lead to environmental degradation in another.

While different social science disciplines have different areas of enquiry the differentiation between these disciplines can never be absolute or complete, as the actions, structures, and processes that different social scientists study can never be isolated from the broader social context. In this respect disciplines and disciplinary differences are always constructed and contestable. What all social science disciplines share is an interest in social change and what happens when people act together, or against one another. Various theoretical concepts cut across several social science disciplines. Three are now examined: inequality, power, and values.

Inequality

People are united and divided by nationality, ethnicity, social class, ownership of wealth, gender, life opportunities, and so on. Inequality is an inescapable feature of the social world that manifests itself in different ways, as several studies in this book have shown. The problem of climate change is in large part one of inequality. As Chapter 10 argues, about 500 million people are responsible for about half of global greenhouse gas emissions, while the poorest 3 billion emit almost nothing. One manifestation of inequality is that poorer districts tend to host waste management sites whereas richer areas do not (Chapter 14). Chapter 8 introduces the concept of environmental racism to explain the tendency by politicians to site toxic waste sites close to minority populations in the United States. The chapter suggests that the pursuit of sustainability requires greater representation in the political process from those marginalized groups who bear a disproportionate share of environmental risks and harms.

Social scientists often point to the role of inequality in driving environmental degradation. A commonly drawn distinction is that between the unsustainability of poverty, which is associated with low per capita incomes and the degradation of natural resources such as land and forests to meet basic needs, and the unsustainability of affluence, which is associated with resource degradation driven by high levels of consumption to satisfy wants (Blowers and Glasbergen, 1995). Those who argue that inequality drives environmental problems stress that addressing social inequalities is not only morally desirable in its own right, it is a necessary precursor to solving environmental problems. Tackling social inequalities requires tackling some fundamental power imbalances in society.

Power

Power is a concept that several authors have used in this book, but upon what is power based? One common theoretical approach is that power is based on resources. An actor has power if they can mobilize resources such as people, technology, finance, and so on. The power of actors comes from the resources they control and can deploy to change the behaviour of other actors. So if actor A has more resources than actor B then the former will exert more influence over the latter. On this view power is relational; some actors may have more or less power than others. For example, those who promote and protect the coal industry in Australia have more power than those who oppose it (Chapter 7). In Africa the unequal access of different social groups to water is an expression of underlying power relations (Chapter 4).

A second theoretical approach to power is that it is the ability to construct and shape discourses. A **discourse** is a body of language that embodies certain understandings and ideas. Discourses regulate the production of meaning in social life; they help to structure how people think and act. An understanding of discourse is thus an important part of the study of environment, development, and sustainability. The most effective discourses are those that are taken for granted as 'common sense'.

There are many examples in this book of discourse that informs academic discussion and policy debate on the environment. For example, the discourse of romanticism has played an important role in shaping the relationship between nature and society in the United

States (Chapter 8). The evolving discourse of sustainability and sustainable development comes through in different guises in different places and for different issues, including sustainable consumption (Chapter 5), sustainable transport (Chapter 6), sustainable forest management (Chapter 11), and sustainable cities (Chapter 18, 21).

These approaches to power—as the control and deployment of resources, and as the construction of discourses—are related. Powerful actors have the material resources that enable them to shape and promote new discourses in society. Actors often frame their interests by drawing from the language and assumptions of commonly accepted discourses in order to legitimize their positions and to provide their statements with authority.

In all countries those who hold political and economic power have the capacity to help shape the law. The law sets the limits to what is considered acceptable or permissible in a society, although to greater or lesser degrees the law also reflects the interests of those who rule. The clearance of slums in Mumbai is an example of how the powerful may use the law to act against the interests of marginal groups (Chapter 9). US law has been used to implement a trade blockade of Cuba, which has affected the economic development and environmental policies of that country (Chapter 6). While most social actors choose to obey the law, some do not if they believe that the law leads to undesirable effects, such as environmental despoilation. As one of the activists quoted in this book argues, 'When the laws are unjust or are destroying our future, people of conscience have a responsibility to act' (Chapter 7). This suggests that there are moral obligations and social values that are important, but which do not find expression in the law.

Values

A social value is something that is considered important to a particular actor, group, or society. Examples include cultural heritage, acceptance of a commonly accepted way of life (democratic values or family values), or belief in certain principles (fairness and justice). Because different actors adhere to different social values, issues such as 'sustainability', 'development', or 'conservation' become matters on which there are, inevitably, different interpretations. Contestation, maybe conflict, is often inescapable. It is common to distinguish between two clusters of environmental values in the social sciences: intrinsic values and instrumental values (Chapter 16). Those who view the environment principally as a resource for economic activity emphasize the instrumental value of the environment which should be conserved as a means to an end, because we can use natural resources in the manufacture of foods, medicines, industrial products, and so on. Other actors stress intrinsic value; the environment matters as an end in itself and has value in its own right.

Many economists equate value with the market worth of a product or service, with value expressed as price or utility. Environmental economists research how environmental goods and services may be provided more efficiently by putting a price on them, or by charging those whose activities cause environmental problems. The London congestion charge, from which low emission vehicles are exempt, is one of the economic instruments deployed under the London Mayor's Climate Change Action Plan (Chapter 18).

Social values are embedded in many of the principles about which you have read in the book. For example, values such as fairness and equality inform the principles of *intra*generational and *inter*generational equity. Intragenerational equity is the principle of fairness over space, that actions should not impose an unfair or undue burden on individuals or groups within the present generation. Intergenerational equity is the principle of fairness over time, that the present generation should not impose unfair or undue burdens on future generations. Intergenerational and intragenerational equity are central themes in most debates on environmental ethics (Chapter 16).

Social scientific analysis is not enough

By analysing the role of social inequalities, power, and values in causing conflict over natural resource use and economic development the social sciences can offer suggestions on how more just and sustainable societies can be created. However, social scientific analysis is not enough. Many environmental problems have their origins in technology, and technological innovations may also serve as a force for sustainability and conservation (Chapters 19 and 20). It is to the study of technology that we now turn.

The dual nature of technology

Humans and other animal species modify and use parts of their environment to their own ends. In its simplest form this can, for example, involve humans creating stone tools by modifying materials found around them and animals using sticks and twigs to get to their food. In doing so they inevitably modify the environment in which they live. Such activities are a simple form of technology. They increase the access an individual has to a resource and allow that resource to be 'exploited' more efficiently. In these cases the technology is built on craft skills without any understanding of the scientific principles behind it, but as science has developed, the development of technology has been based more and more on science.

Individuals learn from each other and, as they do so, 'rules' and general principles (both craft based and scientifically based principles) for making things are disseminated. Rules and principles discovered from earlier technologies are used to develop new ones; and technology thus becomes a collective activity.

Technology, then, is not just about the final artefact that results from a development: it is also about the process involved in its production and use. Its impact is not just about the use of the artefact but also about all the processes, the people, the materials, the organization, the energy, used in the production and use of the artefact.

Technology is also about organizing things in a systematic way. Devising a better technological artefact may not be enough on its own unless you then teach people how to use and maintain it, and what to do when it breaks. This can lead to society itself being transformed. Technology thus not only shapes the artefact, it can also shape the society that uses it. If technology has the power to shape society then technological decisions can become political decisions, as Chapter 18 illustrates with respect to the role of political leadership.

So, technology is much more complex than being just about the artefacts produced. It has been defined in many ways, some broad in their coverage and some not so broad. From the above discussion, I would argue that technology should be defined broadly as 'the application of science and other knowledge to practical tasks by people working in organizations and using machines'.

This definition implies that technology involves interactions between people, their organizations, and what they produce and consume.

The definition and the previous discussion also imply that the interactions between society, technology, and the environment need to be considered to understand how technology affects our lives. These interactions are complex and so it is necessary to look at every issue in a broad, systemic way and to involve all those likely to be affected. The multiple interactions are represented in Figure 26.1.

If we accept this broad approach to technology there is a consequence of which we should be aware. It is that we have to allow into our discussions and analyses the 'messy' subjectivity of human beings and accept the validity of people's values.

The interactions discussed above can lead to technology being a benefit to the well-being of people and the environment or to the detriment of one or both. Examples include the over-exploitation of natural resources in Uganda (Chapter 3), waste management in Eastern Europe (Chapter 5), the use of hand-drilled wells and improved water-lifting techniques in Niger (Chapter 4), tropical forest management (Chapter 11), and controlling climate change (Chapter 10).

These contrasting (good or bad) impacts are often referred to as the *dual nature of technology*, which applies to all levels of technological development, from individual artefacts to industries and to the level of the global economy. The 'dual nature of technology' holds that technology is neither inherently good nor bad. Its impact depends on the use we make of it, but it does bring with it the potential for both benefit and harm.

Many discussions today about the impact of technology on the environment can be quite polarized, especially where new technologies are involved. People are often strongly for or against the introduction of a particular technology. The issues surrounding each case may be quite different but we appear to be confronted with a paradox when we make use of technology. While the introduction, for example, of a new consumer appliance or a new application of existing technology may bring obvious benefits to individuals or society, in turn it nearly always seems to expose us to new risks and dangers.

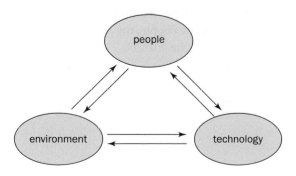

Figure 26.1 Interactions between technology, people, and the environment.

This dual nature arises from the way societies decide to use a particular technology (or not). This makes sense if you take a broad view of technology, outlined above. This is the understanding that technology, and its uses from artefacts to infrastructure, is the product of human and social action. It is a major driver of the development of societies and their economies, but the forms and directions these take are not inevitable.

There are two further concepts which are useful in assessing the possible impact of a given technology on sustainable development. They are the *rebound effect*, and *technological leapfrogging*. The rebound effect points to the tendency of humans to accommodate to change in ways that aren't always intended by the instigators. Examples of this include leaving low-energy lights on for longer, and drivers of cars with safety belts and air bags driving faster. It is a *pessimistic* view of technology. Technological leapfrogging, on the other hand, represents a decidedly *optimistic* view. It points to the possibility of a breakthrough on a global scale to clean, modern technology for countries of the south, without going through many of the stages of heavy industry taken by the north and, hopefully, avoiding the worst of their environmental impacts. (See also Chapter 8 for a discussion of pessimistic and optimistic viewpoints of technology.)

So the final question. Is technology a force for good or evil in our ability to achieve sustainable development? Well, it depends, not only on the technological artefacts and systems developed but also on the will to use them and to adapt our societies to take advantage of the technological solutions.

The scale and range of environmental problems we face are daunting, but problems can be addressed and harm can be reduced. Technology can contribute towards sustainability and development. Whether or not it will in the future depends on how we use it. So whether you are pessimistic or optimistic about the future, how we manage and apply technology should be a central concern of us all.

Conclusion

The three sections above describe three very different approaches to understanding environmental change. Scientific approaches stress robust observation and interpretation of what is happening in terms of underlying mechanisms. For this book, these are the physics of how energy from the sun becomes trapped on earth by increased carbon dioxide concentrations in the atmosphere, the chemistry which reveals the molecular mechanisms of climate change, and the ecology of the ways in which living organisms interact and are interdependent. Social science approaches emphasize understanding processes of social change and how these interact with environmental concerns—social change both leads to environmental concerns (e.g. as in the industrial revolutions) and can be a consequence of those concerns (e.g. attempts to alter lifestyles based on high consumption). Finally, there are two key dimensions to technological approaches. First, they combine and organize the application of scientific knowledge and 'other' knowledge, where much of the latter derives from human experience in the world. This experiential (or 'craft') aspect means in turn that technology relates more to processes of learning than any fixed analytical frameworks. Secondly, technology is relational to people and the environment.

It has the potential both to help meet environmental challenges and to be detrimental to the environment, depending on the use we make of it.

Science, social science, and technology are all important if we are to understand and act to meet environmental challenges. By themselves, however, none can capture the full story of environmental change.

■ REFERENCES

Blowers, A. and P. Glasbergen, P. (1995) The search for sustainable development. In P. Glasbergen and A. Blowers (eds) *Environmental Policy in an International Context: Perspectives on Environmental Problems*, pp. 163–183. London: Arnold.

■ SOURCES

Open University (2005) Conclusion. *T172 Working with our environment: technology for a sustainable future*, pp. 10–11. Milton Keynes: The Open University.

Open University (2005) Introduction. *T172 Working with our environment: technology for a sustainable future*, pp. 5 and 36. Milton Keynes: The Open University.

27

Conclusion

Gordon Wilson

The many perspectives on environment, development and sustainability are a resource for us to learn from, gain knowledge, and thereby, act appropriately. A major challenge, however, is to work within, while simultaneously challenging, the potentially negative dimensions of difference in terms of inequality and power relations.

The above formed the fourth and final premise for this book which was outlined in Chapter 1. As a result, rather than seek and adopt a single set of unifying ideas about environment, development, and sustainability, the book has deliberately set out to record and embrace difference, both:

- within chapters where authors contrast situations, and concerns and priorities of different actors, and
- between chapters and the different perspectives the authors have brought to bear.

By way of drawing together the book, this last chapter engages with the final proposition, which is to work with difference, and to use the very fact of difference to further our understanding and inform policy and action on environment, development, and sustainability.

First, however, we need to find a way of making sense of the myriad of ways in which the issues and perspectives have been represented by the different authors.

Making sense of difference

Making sense of anything usually starts by creating an order through grouping individual elements within higher-level categories and making links between them. For example a basket of food items might be grouped under the following categories: vegetables, meat, and dairy; or under fresh and processed food. In this vein, a basic approach for examining different perspectives on our environment, is to divide them into intrinsic and instrumental ways of valuing it. Chapters 16 and 26 expand on this distinction at a conceptual level, but basically, when we value the environment for its intrinsic worth, we are valuing it for itself, as something 'natural', which we then seek to conserve. When we value its instrumentality, however, we do so in terms of its practical use for human beings.

Differences arise through the emphases we put on each of these ways of valuing the environment. This is illustrated in this book by Ben Crow (Chapter 8) where he contrasts one view in the United States of reverence for wilderness, and another which is more socially situated and concerns the dumping of toxic waste arising from the instrumental use of the environment. Dina Abbott (Chapter 9) makes a similar point when she contrasts the desire of Mumbai's richer residents for open, wild space, while its slumdwellers view the environment in public health terms. Intrinsic values also surface in Roger Wheater's description of national parks in Scotland (Chapter 12) and Kevin Winter's (Chapter 24) account of collective action to maintain natural indigenous vegetation in the Betty's Bay area of South Africa.

One should note, however, that intrinsic values do not necessarily predominate solely among richer, socially privileged people as might be implied from a casual glance at these chapters. Although not featured as a chapter in this book, the national park movement in England and Wales in the 1930s was at least in part a working class movement to gain access to the wild uplands that lay between industrial cities. These uplands were highly valued as an escape from the drudgery and alienation of the factories.

Thus, the real world is messy and even in situations where intrinsic values are apparently to the fore there is often an overlapping instrumentality in terms of human needs. National parks, wild open spaces in cities, and community organization to preserve indigenous vegetation are all responses to these needs which involve managed preservation of a 'natural' environment. Although such needs are unlikely to be seen directly in material terms they can nevertheless be conceived as developmental.

While accepting the real-world messiness, the intrinsic–instrumental distinction is useful for both thinking about the perspectives of different people and of thinking about one's own perspective, as Martin Reynolds argues in Chapter 16. Most authors of this book, however, have assumed a more explicitly instrumental value for the environment, accepting that it is a material resource for 'development' (first premise, Chapter 1). This is true whether the 'development' relates to rich 'developed' countries or to poorer 'developing' ones.

Consequently, most chapters tend to consider environmental change in relation to different views of social and economic development and of how both environmental resources and development pathways can be sustained. Generally speaking these authors qualify more or less explicitly the view that development is, or should be, simply a matter of economic growth. Whatever the other arguments, at a global level, unbridled economic growth is probably unsustainable environmentally (Chapter 15), particularly when the emergence of China as an economic power (Chapter 2) is taken into account.

What we have then, at global, regional, national, and local scales, is an array of differences in which environment and its relation to development and sustainability are constructed. We might categorize such differences under the following headings:

• *Environment as a source of livelihood.* In economically poor countries where the population is still predominantly rural, for example in Uganda (Chapters 3 and 12), Niger and Ethiopia (Chapter 4), Kenya (Chapter 19), and Zimbabwe (Chapter 23), the environment is a more direct source of livelihoods than in richer, urban-based countries. Insofar as development concerns sustaining and improving rural livelihoods, several of these chapters focus on the conflicts surrounding how environmental resources are deployed, and of whose development they are serving. The one chapter (13) which concerns direct use of an environmental resource for livelihoods in a rich-region context—along the River Rhine in Western Europe—also interestingly focuses on conflicts over its various uses and of how they are managed across several countries.

- *Environmental constructions based on social difference*. We have already noted the different constructions between the poor and rich in Mumbai, and between middle class and poor ethnic minority communities in the United States. Chapter 14 also contrasts the environmental constructions around waste management in relatively poor Kampala (Uganda)—where it constitutes a public health concern—and relatively affluent Birmingham (UK), where the concerns are around waste minimization and recycling.

- *Sustainability constructions arising out of political circumstance*. In Chapter 5, Petr Jehlička argues that people in Eastern European countries have long practised waste minimization and a high degree of self-sufficiency, practices which are at odds with sophisticated 'end-of-pipe' technological solutions to waste management in the European Union. These East European practices are a legacy of the Soviet bloc of which these countries were part from the end of the Second World War until the last decade of the twentieth century. Waste minimization was a response to the austerity of these years, where as much as possible was reused, while food self-provisioning allowed a measure of independence from totalitarian rule. The collapse of the Soviet Union also had a big impact on communist Cuba because it formed the latter's main export market. James Warren argues in Chapter 6 that this, combined with the ongoing United States blockade of Cuba, has resulted in country-wide austerity, yet a transport system which, in contrast to transport in western capitalist societies, is more sustainable because of its lower reliance on the private car.

While the above suggests diverse constructions of environment, development, and sustainability within and between places, based on livelihood opportunities, social differences, and political circumstance, further constructions can be seen to arise from different kinds of professional knowledge of the book authors themselves, which don't necessarily belong to any particular place. Chapter 26 highlighted one way of categorizing these different knowledges, distinguishing between science, social science, and technological backgrounds. They are examined briefly below to draw out their respective emphases, but note that such generalizations must be qualified as nobody fits their categorization exactly and one should avoid dangers of stereotyping.

- *Science background*. These chapter authors have examined the interactions of the human world with the physical and non-human biological worlds (respectively Chapters 10 and 11). They conclude that there is a problem. The great cycles—the carbon and hydrological cycles—and the ecosystems comprise interdependent elements, and there is also interdependence between these cycles. Human activity is running roughshod over this delicate interdependence, particularly in relation to the carbon cycle (with knock-on effects for the hydrological cycle and ecosystems) where it is pushing the world to a tipping point through climate change (Chapter 10). There is an inescapable logic, and hence determinism, to the scientific analysis—human beings must take actions to restore the interdependent balances (Smith *et al.* 2007). The scientists tend to frame action, therefore, in terms of goals which need to be achieved, these goals having been determined through scientific analysis.

- *Social science background.* These chapter authors have focused more on the relationships of people and social groups with the environment. They thus provide reasons why people relate to the environment in particular ways, as noted above. Social scientists also analyse how science can be mobilized to support economic and social change, as has been in the case of establishing tissue-culture bananas for commercial production in rural Kenya (Chapter 19).

- *Technological background.* The authors who would describe themselves in this way have generally framed the issues in terms of what is practically possible. This often involves an assessment of technical possibility and desirability in particular social, economic, and political contexts. Thus, Chapter 4 concerns enabling the farmers in Niger and Ethiopia to build on their local knowledge to make incremental improvement to water supply and irrigation. Chapter 18 tells the story of policies to make London carbon-neutral within a particular political context. Chapter 20 argues that sustainability on a large scale requires a design-systems approach, which integrates the environmental, social, and economic dimensions of sustainability. Chapter 21 describes tools that can be used for integrating these dimensions and working holistically in a specifically private sector context, while Chapter 25 focuses more generally on the tools that are necessary to help make decisions that incorporate environmental as well as economic, social, and other concerns.

Another way of dividing the author perspectives is between those who are academics and work in universities, and those who are practitioners, working practically towards a goal, however distant, of sustainable development. Although there are overlaps in the sense that many of the practitioner authors also have strong academic qualifications, it is quite easy to tell them apart. The practitioners obviously think of themselves as, and have a sense of pride in being, active agents in the processes, and their chapters tend to promote ways of working in which they themselves engage. Chapters 22 and 23, which describe partnership and participation approaches in Zimbabwe and are written by practitioners of the regional NGO, Environment Africa, are a classic example. So too is Chapter 7 which reveals how an environmental movement in Australia works.

In contrast, the academic authors tend to think about action as something separate from themselves, where they stand back in order to analyse. Examples include Chapter 9 on different actions in Mumbai, Chapter 13 on conflicts over environmental resources on the River Rhine, Chapter 17 on how sustainable development policy and action is constructed in Wales, and Chapter 24 on how people came to act collectively on environmental initiatives in Betty's Bay, South Africa.

Both practitioner and academic perspectives have advantages. The practitioners' contribution is their ability to provide lived insights and experiences of what it means to act. The academics' contribution is the insights to be gained from standing back, conceptualizing (and thus helping provide analytical tools for others to use), and bringing a more generalized knowledge to bear on situations.

Difference as a resource, and interdependence

Despite having put the perspectives and cases of previous chapters in some semblance of thematic order, the overwhelming picture is of a complex diversity (third premise, Chapter 1). Can the picture be made to hang together? Chapter 1, and the book's fourth premise repeated above, suggested that, rather than worry about difference, we should embrace it as a resource for learning and constructing knowledge through joint engagement about environment, development, and sustainability and the actions we should take. Rather than an end point, sustainability then becomes a process of continuous learning for **innovation** in its general sense as knowledge put to productive use (Johnson and Wilson 1999), a view also put with respect to attempts to design and construct eco-cities at the end of Chapter 21. This view also accords with sustainability's dimension of robustness—the ability to continue over time (second premise, Chapter 1). Such robustness might partly be about pre-empting shocks through learning and taking action, but it is equally about learning to cope with them and take adaptive action when they do occur.

The point about difference being a resource for learning is not to try and make us all the same, but (to draw a parallel with the sustainable use of physical resources) to use that difference productively while sustaining it for further learning. This requires, first, respect for each other and our different knowledges, but also acceptance that what we learn together and evolve will always be mediated by our prior knowledge which is produced by our personal experiences and histories. This is implied in Chapter 4 which argues for building on existing knowledge to make incremental improvements to water supply and irrigation in rural Niger and Ethiopia. It is, however, important not to fall into the essentialist trap of romanticizing local knowledge as being the only credible knowledge. When the authors of Chapter 23, for example, recount the moment that they realized that the champion farmer in Zimbabwe had come to understand the principles of sustainable development the 'African way', we should understand the comment in terms of the farmer mediating new knowledge through what he knows already, rather than essentialist notions that Africans are unchanging due to some primordial traits.

Difference as a resource for learning through joint engagement is the optimistic side of the coin. On the other side, difference between people and social groups is equated with power, inequality, and injustice. Chapter 15 points to the growing social inequalities under processes of **globalization**. As already noted, at local scales the book has witnessed inequalities and power struggles between poor and rich in Mumbai (Chapter 9) and between tropical forest dwellers and their government and big private corporations in Uganda (Chapters 3 and 12). Meanwhile, at country scales, we have seen that, despite its economic growth, people in China are poorer than many other parts of the world (Chapter 2), that Eastern Europe is having to take on environmental agendas of Western Europe in order to gain accession to the European Union (Chapter 5), and that Cuba is locked in a power struggle with the United States (Chapter 6).

Writing from a social science perspective, David Humphreys in Chapter 26 also points to how the most powerful actors influence constructions of knowledge so that they represent, and perpetuate, their interests. Such constructions can then become internalized by everybody and appear as 'common sense'. This critical perspective on how knowledge is constructed is also often applied to participatory and partnership approaches to action (Cooke and Kothari

2001; Hickey and Mohan 2004) that are promoted in several chapters of the book. The point is that power differences do not simply melt away because of the label 'participation'. Adeline Muheebwa's call for inclusiveness of women and youth in community forest management in Uganda (Chapter 12) is an implicit acknowledgement of this critique.

Taking on board both the optimistic and pessimistic perspectives on difference presents a conundrum. There are two possible ways out, both of which invoke the concept of interdependence:

1. Establishing interdependence through existing common practical interests. Most likely, such interdependence and common interest will have emerged in particular places and their particular histories. It is usually observed at a local scale, where there is a relatively high degree of social cohesion. The best-documented example in this book is the collective action of relatively privileged white South Africans in Betty's Bay for preserving the indigenous flora (Chapter 24).

2. More commonly, and from local to global scales, is to establish interdependence out of our differences, even when these are also a function of social and political inequalities. One clue here is that nobody has a monopoly on power, and there is always room for negotiation. The very term 'interdependence' recognizes this and implies a certain mutuality in terms of meeting the practical interests of different groups. Thus:

   • At a local scale, Chapter 9 suggests that the rich and poor of Mumbai are in fact interdependent which gives the latter some negotiating power. Chapter 22 refers to mutuality as 'win–wins', in this case between the private sector, local government, local community, and an environmental NGO over the state of the river which runs through Harare, the capital of Zimbabwe.

   • At a country scale, Chapter 17 describes the ways in which the Welsh Assembly Government promotes engagement between multiple stakeholders from both inside and outside government in policy development and action for sustainable development. Alan Thomas thus refers to a process of governance rather than government, where plural interests are accommodated. The broad governance structures for National Parks in Scotland described in Chapter 12, operate also with a range of stakeholders.

   • At a global scale, Chapter 15 suggests that global inequality, professional insecurity, and the impacts of climate change might combine to bring to a halt the globalization process.

   • Working across scales, Chapter 19 examines how a tacit (if possibly temporary) agreement emerged on the need to develop tissue-culture bananas in rural Kenya—involving scientists, government officials, international aid donors, and local farmers.

The above two ways of viewing interdependence rest on either already having in place evolved, common, practical interests or consciously negotiating collective ways of meeting the diverse interests of social groups. Being organized around parochial, practical interests inevitably raises the question, however, whether such actions are sufficient to meet general environmental challenges such as climate change which don't necessarily map onto immediate interests. In themselves they are undoubtedly insufficient, but they are nevertheless the starting point for changing the only thing we can change, which is ourselves (Weick 1995, 2001). Rather than seeing negotiations to meet diverse practical interests as boundaried activities, therefore, they can be viewed as starting points for working together and establishing trust. Mutual trust is a prerequisite for deeper communication and hence joint learning because it helps us to expose ourselves and our ideas, and be open with one another. The German philosopher Jürgen Habermas (1990) refers to joint learning in this way as 'communicative action'.

As noted above, Smith *et al.* (2007) point to the contribution of science in having established the 'fact' of interdependence within the non-human world. This is a crucial general contribution to our factual understanding over and above the specific analyses provided by science of, for example, climate change or biodiversity loss. Smith and colleagues also argue, however, that, at the human level, interdependence has to be seen as much more emergent, achieved in the present through negotiations of interests and meanings. This conclusion is also borne out by the chapters of this book and the analysis above. By bringing together such a variety of authors, themes, and situations under one volume the book as a whole has also contributed in a small way to this emergent interdependence.

▓ SUMMARY

- A basic way of categorizing difference in relation to environment, development, and sustainability is through the relative emphasis put on the environment's intrinsic and instrumental value.

- This book has highlighted subcategories with respect to the instrumental use of the environment for economic and social development. These subcategories include constructions of environment in terms of different livelihood opportunities, social difference, and political circumstance.

- Different approaches to environment, development, and sustainability can also be categorized according to professional perspectives—this book has been authored by scientists, social scientists and technologists, and academics and practitioners.

- If sustainability is conceived as a process of joint learning and consequent action, difference between people and groups is a valuable resource. As a downside, difference is also associated with inequality and more/less powerful groups. The concept of interdependence, however, implies that nobody's power can be absolute and there is room for negotiation of interests leading to joint learning for sustainability.

▓ REFERENCES

Cooke, B. and Kothari, U. (eds) (2001) *Participation: The New Tyranny?* London: Zed Books.

Habermas, J. (1990) *Moral Consciousness and Communicative Action.* Cambridge: Polity.

Johnson, H. and Wilson, G. (1999) Institutional sustainability as learning. *Development in Practice* 9(1): 43–55.

Hickey, S. and Mohan, G. (eds) (2004) *Participation: From Tyranny to Transformation.* London: Zed Books.

Smith, J., Clark, N. and Yusoff, K. (2007) Interdependence. *Geography Compass* 1(3): 340–359, 10.1111/j.1749–8198.2007.00015.x

Weick, K. (1995) *Sensemaking in Organizations.* London, Thousand Oaks, New Delhi: Sage.

Weick K. (2001) *Making Sense of the Organization.* London, Thousand Oaks, New Delhi: Sage.

GLOSSARY

Note that this is *not* a comprehensive list of the key terms and concepts used in the book. Most key terms and concepts are explained, at least briefly, in the chapters in which they appear, sometimes within the main text and sometimes as a separate short section of highlighted text.

The criteria for further inclusion in a glossary are twofold:

1. Those concepts and terms which appear in more than one chapter, in order to save repeating in-chapter explanations. Exceptions to this 'rule' are the underlying concepts of environment, development, sustainability, and sustainable development. These are explored in, and form the basis of, Chapter 1.

2. Concepts and terms where we consider it advisable to expand on the meanings provided within the chapters.

The glossary concepts and terms have been highlighted in bold type where they first appear in a chapter, and also below where one definition refers to another definition.

B

Biodiversity A word formed by the contraction of the term 'biological diversity' that describes the variety of life on Earth. It usually refers to species diversity, that is, the number of different plants and animals of all types that are found in a given **ecosystem.**

C

Civil society is that area of organized life in society which is independent of the state, the market-based institutions of business and commerce and the private lives of our families. Typically in this book, it is used as a catch-all phrase to describe non-governmental organizations (especially those involved in development work) and organized, membership-based community groups. A strong civil society is considered important for a flourishing democracy.

Corporate social responsibility (CSR) is where business organizations consider the interests of society by taking responsibility for the impact of their activities on customers, suppliers, employees, shareholders, communities, and other stakeholders, as well as the environment. This obligation is seen to extend beyond the statutory obligation to comply with legislation and sees organizations voluntarily taking further steps to improve the quality of life for employees and their families as well as for the local community and society at large. (Adapted from Wikipedia http://en.wikipedia.org/wiki/Wikipedia.)

D

Discourse A discourse is a way of communicating about an issue through language (the phrases and vocabulary we use). It represents a particular way of thinking about an issue(s). To do so, it embodies certain understandings and ideas. Discourses thus regulate the production of meaning in social life; they help to structure how people think and act. The most effective discourses are those that are taken for granted as common sense. (Adapted from Chapters 8 and 26.)

E

Ecosystem describes a community of living organisms and their interactions with one another, plus the environment in which they live and also interact. Thus it includes both living and non-living components and the links between them. The term can be applied at many scales, e.g. a pond, rainforest, oceans, or even the Earth itself.

G

Globalization is the process of increasing interconnectedness of the world, facilitated by modern communications technology and transport, and, in the realm of the economy, by the ability to produce different parts of the same product in different places. Optimistic views on globalization claim that this economic integration is overall a good thing which ultimately allows everybody to enjoy the fruits of market-based capitalism. Pessimistic views, however, argue that increasing integration reinforces existing global inequalities and power relations. A third view, the transformationalist view, argues that globalization is creating a new world order, where new concentrations of power and inequality, wealth and poverty, are forming in both the hitherto rich countries and the historically poor countries. (Adapted from McGrew 2000.)

Governance is defined in various overlapping ways. Stoker (2004) describes it as the conditions for ordered rule, which involves a variety of actors from government, the business and commercial sector, and **civil society**. The Commission on Global Governance (1995: 2) defines it as 'the sum of the many ways institutions, public and private, manage their common affairs'. Common to these definitions is the idea that more than 'government' is required to manage a society effectively. The term **good governance** is also often used, and is the sound management of a country's economic and social resources (Potter 2000: 382). In this book, good governance is used to imply inclusion of a full range of stakeholders in decision making.

Gross domestic product (GDP) is the market value, usually measured in United States dollars, of all goods and services produced in a country over a year. It is commonly used to express the size of a country's economy. A common refinement is to adjust the value of GDP for **purchasing power parity (PPP)**. GDP per capita is the total value divided by the population of the country, and so gives an indication of the economic output per head of population. A country like China might have a very high total annual GDP because of the huge size of its population, but on a per capita basis it remains poor.

I

Innovation In a narrow sense, innovation means the process by which new goods and processes are developed and brought to market. More broadly in this book it refers also to new ways of organizing and doing things in social life. Innovation depends ultimately on learning and creating new knowledge. The definition used in Chapter 19 therefore refers to it as knowledge put to productive use.

Institutions can be organizations or established aspects of wider society which embody common values and ways of behaving. An example of the former is the UK National Health Service with its value of free health care; an example of the latter is the institution of marriage. **Institutional development/change** concerns changing these values and ways of behaving. It describes what is required to spread new values and ways of behaving throughout organizations and embed them into management practices in all sectors. This means change not only in organizational structures, processes, goals, and mission statements, but also in the mindset which determines informal ways of working, behaving and justifying decisions. (Adapted from Chapter 17.)

L

Life cycle assessment (LCA) also known as life cycle analysis, life cycle management and 'cradle to grave' analysis. LCA traces environmental impacts at all stages of a product's 'life cycle'. This starts from the impacts arising from extracting and processing raw materials, followed by impacts from manufacturing and delivering a product, impacts in using it and, finally, what happens at the end of a product's life e.g. whether it is reused, recycled, or disposed of.

M

Millennium Development Goals (MDGs) are eight goals established by the United Nations and agreed internationally to be achieved by the year 2015. They encapsulate the world's main development challenges:

1. Eradicate extreme poverty and hunger through (a) reducing by half the proportion of people living on less than one US dollar a day; (b) achieving full and productive employment and decent work for all, including women and young people; (c) reducing by half the number of people who suffer from hunger.

2. Achieve universal primary education by ensuring that all boys and girls complete a full course of primary schooling.

3. Promote gender equality and empower women through eliminating gender disparity in education at all levels by 2015.

4. Reduce child mortality through reducing by two thirds the mortality rate among children under 5.

5. Improve maternal health by (a) reducing by three-quarters the maternal mortality ratio; (b) achieving, by 2015, universal access to maternal health care.

6. Combat HIV/AIDS, malaria, and other diseases through (a) halting and beginning to reverse the spread of HIV/AIDS; (b) achieving, by 2010, universal access to treatment for HIV/AIDS for all those who need it; (c) halting and beginning to reverse the incidence of malaria and other infectious diseases.

7. Ensure environmental sustainability through (a) integrating the principles of sustainable development into country policies and programmes; and reverse loss of environmental resources; (b) reducing biodiversity loss, achieving, by 2010, a significant reduction in the rate of loss; (c) reducing by half the proportion of people without sustainable access to safe drinking water and basic sanitation; (d) achieving significant improvement in the lives of at least 100 million slum dwellers, by 2020.

8. Develop a global partnership for development through (a) developing further an open, rule-based, predictable, non-discriminatory trading and financial system; (b) addressing the special needs of the least developed countries.

P

Poverty is defined in this book in material terms where its measurement is based on the level of household expenditure. The World Bank has defined 'absolute' poverty as living on less than US$1.00 per day in **purchasing power parity (PPP)** terms. This definition is thought by others, however, to be too narrow. A broader approach defines poverty as having multiple dimensions, such as low income, poor literacy, poor health, poor access to clean water and sanitation, and so on. A third approach defines poverty in terms of social inclusion—the ability to play a full part in the society in which one lives.

Purchasing power parity (PPP) is a value put on goods and services in different countries (and in this book **GDP**) which has been adjusted to account for different costs of living in the countries. One reason for making a PPP adjustment is that non-traded goods tend to be cheaper in developing countries, and so a given currency will go further in some countries compared with others.

U

UNCED The United Nations Conference on Environment and Development, informally known as the Earth Summit, was held in Rio de Janeiro, Brazil in 1992. This major international conference was very productive in terms of the number and scope of its outputs. From UNCED emerged the Rio Declaration on Environment and Development, Framework Convention on Climate Change, Convention on Biological Diversity, and Agenda 21, which set an extensive international agenda for action for sustainable development for the twenty-first century.

The Rio Declaration on Environment and Development was a statement of 27 key principles for sustainable development and international law including the precautionary principle, the polluter pays principle, public participation, and environmental assessment. (Adapted from Open University 2006.)

W

WSSD The World Summit on Sustainable Development, convened by the United Nations, was held in Johannesburg, South Africa in 2002, 10 years after **UNCED**. From WSSD came reaffirmed commitment to sustainable development in the Johannesburg Declaration which committed to 'a collective responsibility to advance and strengthen the interdependent and mutually reinforcing pillars of sustainable development – economic development, social development and environmental protection – at local, national and global levels'. (Adapted from Open University 2006).

▨ REFERENCES

Commission on Global Governance (1995) *Our Global Neighbourhood: The Report of the Commission on Global Governance.* Oxford: Oxford University Press.

McGrew, A. (2000) Sustainable globalization. In Allen, T. and Thomas, A. (eds), *Poverty and development into the 21st century.* Oxford: Oxford University Press.

Open University (2006) T863 *Environmental decision making: a systems approach.* Milton Keynes: The Open University.

Potter, D. (2000) Democratisation, 'good governance' and development. In Allen, T. and Thomas, A. (eds) *Poverty and Development into the 21st Century.* Oxford: Oxford University Press.

Stoker, G. (2004) *Transforming Local Governance.* London: Palgrave Macmillan.

■ INDEX